Bile Acids
Toxicology and Bioactivity

Issues in Toxicology

Series Editors

Professor Diana Anderson, *University of Bradford, UK*
Dr Michael D Waters, *National Institute of Environmental Health Science,*
N Carolina, USA
Dr Timothy C Marrs, *Edentox Associates, Kent, UK*

This Series is devoted to coverage of modern toxicology and assessment of risk and is responding to the resurgence in interest in these areas of scientific investigation.

Ideal as a reference and guide to investigations in the biomedical, biochemical and pharmaceutical sciences at the graduate and post graduate level.

Titles in the Series:

Bile Acids: Toxicology and Bioactivity
Edited by Gareth Jenkins, *Institute of Life Science, Swansea School of Medicine, Swansea University, Swansea, UK* and Laura J. Hardie, *Molecular Epidemiology Unit, University of Leeds, Leeds, UK*

Cytochrome P450: Role in the Metabolism and Toxicity of Drugs and other Xenobiotics
Edited by Costas Ioannides, *University of Surrey, Guildford, UK*

Hair in Toxicology: An Important Bio-Monitor
Edited by Desmond John Tobin, *University of Bradford, Bradford, UK*

Male-mediated Developmental Toxicity
Edited by Diana Anderson and Martin H Brinkworth, *University of Bradford, Bradford, UK*

Visit our website at www.rsc.org/issuesintoxicology

For further information please contact:
Sales and Customer Care, Royal Society of Chemistry, Thomas Graham House, Science Park, Milton Road, Cambridge, CB4 0WF, UK
Telephone +44 (0)1223 432360, Fax: +44 (0)1223 426017, Email: sales@rsc.org

Bile Acids
Toxicology and Bioactivity

Edited by

Gareth Jenkins
Institute of Life Science, Swansea School of Medicine, Swansea University, Swansea, UK

Laura J. Hardie
Molecular Epidemiology Unit, University of Leeds, Leeds, UK

RSCPublishing

ISBN: 978-0-85404-846-5

A catalogue record for this book is available from the British Library

© Royal Society of Chemistry, 2008

Published by The Royal Society of Chemistry,
Thomas Graham House, Science Park, Milton Road,
Cambridge CB4 0WF, UK

Registered Charity Number 207890

For further information see our web site at www.rsc.org

Preface

We present here a full and detailed set of reviews focusing on the toxicology and bioactivity of bile acids. We have brought together world experts in their own fields to discuss the contribution of bile acids to various disease pathologies, as well as discussing the mechanisms behind their activity.

This exciting manuscript, for the first time, details the plethora of biological activities of this fascinating group of naturally occurring chemicals. We hope that this provides a one-stop reference for scientists to gain a fuller understanding of bile-acid activity and function.

Bile acids are essential agents involved in lipid digestion and absorption in mammals. However, they play wide-ranging roles in a variety of disease states ranging from diabetes to cancer. They have evolved exquisite mechanisms for controlling their own synthesis and to ensure that they are produced at correct concentrations and also kept in the correct anatomical environment. It is only when these fine levels of controls are breached that bile acids become associated with disease. This breaching of control mechanisms can occur through dietary means (*e.g.* in obesity) whereby excessive levels of bile acids are produced and converted (via the bacterial flora) to damaging secondary bile acids. Furthermore, lack of reabsorption of Bile acids can lead to liver pathologies. The atypical movement of Bile acids into the oesophagus, stimulated by episodes of reflux, is linked to oesophageal cancer.

We would like to thank everyone who has contributed to this book and hope that the readers of this manuscript find the text stimulating and rewarding in terms of their understanding of the mechanisms underlying bile-acid induced pathology.

Gareth Jenkins and Laura J. Hardie

Issues in Toxicology
Bile Acids: Toxicology and Bioactivity
Edited by Gareth Jenkins and Laura J. Hardie
© Royal Society of Chemistry, 2008

"By the gods, you can swallow your own bile till it kills you"

William Shakespeare
Brutus to Cassius in *Julius Caesar*

In Shakespeare's time, the traditional model of illness was as an imbalance of the four humours: melancholy (or black bile), choler (or yellow bile), blood, and phlegm.

Contents

Issues in Toxicology
Bile Acids: Toxicology and Bioactivity
Edited by Gareth Jenkins and Laura J. Hardie
© Royal Society of Chemistry, 2008

**Chapter 3   Bile-Acid Induction of Apoptosis in Relation to
Gastrointestinal Cancer**
*Katerina Dvorak, Harris Bernstein, Claire M. Payne,
Carol Bernstein and Harinder Garewal*

An Overview of Bile-Acid Synthesis, Chemistry and Function

DENNIS STAMP[1] AND GARETH JENKINS[2*]

[1] Retired Researcher (Dept of Nutrition, University of Toronto), Current address: 23 Fairmar Ave, Toronto, ON, M8Y2C7, Canada; [2] School of Medicine, Swansea University, Swansea SA28PP, UK

1.1 The Bile Acids

Bile acids (BAs) are a group of water-soluble steroids formed during the catabolism of cholesterol, and synthesised in the hepatocytes of the liver. The products, cholic acid (CA), and chenodeoxycholic acid (CDCA), are called primary bile acids. Figure 1.1 shows an overview of the pathways involved in these reactions. These primary BAs are then conjugated, mainly to either glycine or taurine. The conjugated BAs play a pivotal role in fat (and fat-soluble vitamin) digestion and absorption, reaching the colon via the gallbladder, bile duct, and duodenum. BAs are strongly cytotoxic, and are able to act as nuclear sensors, detecting and controlling their own concentrations within the body. Bile acids also play a major role in carcinogenesis of some tissues (liver, gallbladder, upper and lower GI tract). These roles will be described in the following pages and following chapters. BAs are stored in the gallbladder under extremely high concentration (>300 mM), achieved by a constant removal of water and electrolytes. About 5% of these bile acids go to the colon for excretion in the faeces,

*corresponding author

Issues in Toxicology
Bile Acids: Toxicology and Bioactivity
Edited by Gareth Jenkins and Laura J. Hardie

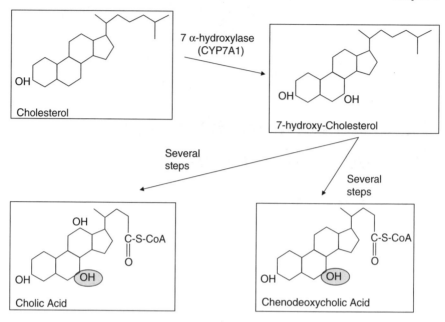

Figure 1.1 The classic pathway for the conversion of cholesterol into the primary bile acids CA and CDCA, involving the 7 α-hydroxylase enzyme (also known as CYP7A1). Simplified from Dr John Chiang.[1] The 7 OH group is highlighted with the shaded circle. This group is cleaved to produce the secondary BAs DCA and LCA.

and since cholesterol is a precursor of BA, this is the only time cholesterol is excreted from the body (as bile). Also present in bile are:

(1) Bilirubin and other pigments resulting from haem catabolism,
(2) Heavy metals such as copper or iron, in excess of bodily needs, and
(3) Lipophilic steroids and drug metabolites that would be insoluble in the urine.

In the colon, deconjugation of the conjugated primary bile acids occurs *via* the action of bacterial enzymes, producing free bile acids. Furthermore, the enzymatic action of the bacterial flora converts the bile acids into secondary BAs, by removing the hydroxyl group from the 7th carbon atom on the molecule. The specific enzyme responsible is 7 alpha-dehydroxylase, which forms deoxycholic acid (DCA) from cholic acid, and lithocholic acid (LCA) from chenodeoxycholic acid. These secondary bile acids then pass into the portal vein and reach the liver, where they join new primary BAs, they are then reconjugated to glycine or taurine in the canaliculi of the liver, and are then stored in the gallbladder. This recycling of bile acids is known as the enterohepatic circulation and can occur 10 times every day. Transport across the canalicular membrane of the liver, is an

ATP-dependent process, aided by the bile-salt excretion pump (BSEP) expression in the canalicular membrane. Conjugation increases the aqueous solubility of the bile acids, and renders these bile acids largely impermeable to the cell membranes of the intestine and duodenum; hence, they are unable to leave the intestinal lumen. This allows bile-acid levels to rise in the lumen, ultimately reaching sufficient concentrations to form micelles, which allow lipid emulsification and subsequent absorption.

Many other BAs are formed at lower levels both in the colon and liver by the bacterial flora and the conjugation with other biomolecules, but this chapter will focus on the more common bile acids; cholic and chenodeoxycholic acids (primary BAs), deoxycholic acid and lithocholic acids (secondary BAs), and their glycine and taurine conjugates. These are the main sub-types of bile acids, as seen in Table 1.1. There are some "minor" BAs that have significant importance. One is ursodeoxycholic acid (UDCA), which, as its name suggests, is abundant in bears, and much prized in Eastern medicine. Human bacterial flora can produce it as well, along with dozens of other BAs and their many isomers. Ursodeoxycholic acid plays a role in human cholesterol regulation, and its medical applications include dissolving gallstones and protecting cells from the harmful effects of other BAs like DCA in cholestatic diseases. When used medically, UDCA is not obtained from bears, but is synthesised from cholic acid, a byproduct obtained from the abattoir.

Table 1.1 Some of the biochemical properties of bile acids.

Bile acid	Water solubility	CMC (mM)	CMpH	pKa	% in bile
Free bile acids					
CA	273 um	13	6.65	5.2	Trace
DCA	28 "	10	7.08	6.2	Trace
CDCA	27 "	9	7.2	6.2	Trace
Glycine conjugates					
GCA	32 "	12	–	3.8	30
GDCA	6 "	6	–	4.8	15
GCDCA	7 "	6	–	4.3	30
Taurine conjugates					
TCA	Very sol	10	–	<2	10
TDCA	Very sol	6	–	<2	10
TCDCA	Very sol	7	–	<2	5
References	(12)	(12,13)	(12)	(14)	(15)

NB: In this table, BAs are divided into 3 groups: Free BAs, Glycine conjugates, and Taurine conjugates. CA = cholic acid, DCA = deoxycholic acid, and CDCA = chenodeoxycholic acid. The values quoted above represent human bile, and were taken from multiple sources. The amount of conjugated and free bile acids in bile is quite variable. Values in this table were determined for single BAs. In actuality, BAs exist as mixtures, and since they are detergents, they will influence each other's solubility characteristics. For example, taurine conjugates are strong sulfonic acids, capable of protonating other bile acids, and thus allowing them to enter the epithelium without any regards for established solubilities.

1.2   Conjugated Bile-Acid Biosynthesis

Figure 1.1 illustrates a condensed version of the classical pathway of bile-acid synthesis, a series of 12 enzymatic reactions that convert cholesterol, which is insoluble, into BAs, which are water soluble. The cholesterol is first converted to 7 alpha-hydroxy cholesterol, followed by the series of enzymatic transformations, eventually producing cholic and chenodeoxycholic acids (not all steps shown). The rate-limiting enzyme in this pathway is cholesterol 7 alpha-hydroxylase (CYP 7A1), which originates from microsomal cytochrome P-450 enzymes, expressed only in the liver hepatocytes.

Another indirect pathway (not shown in Figure 1.1) involves cholesterol reacting enzymatically with CYP 27A1, producing both 27-hydroxycholesterol and 3 beta-hydroxy-5-cholestanoic acid (omitted from the diagram for simplification). This is followed by a series of reactions, ending in the production of chenodeoxycholic acid. The inner mitochondrial membranes are the main reaction site for this pathway. In the adrenal glands, steroid acute response protein (StAR) delivers cholesterol to the mitochondrial membrane. StAR is necessary for steroidogenesis, and thus may provide a reliable source of cholesterol for these reactions.

Another pathway of some importance occurs in the brain; this is the cholesterol 24-hydroxylase pathway. About 25% of the body's cholesterol exists in the plasma membranes of myelin sheaths. Here, the blood–brain barrier prevents cholesterol exchanges with the circulating lipoproteins, which makes it difficult for cholesterol to leave the brain. The cytochrome P-450 enzymes (CYP 46), expressed almost exclusively in the endoplasmic reticula of the brain, allows formation of 24-hydroxycholesterol.

It is impossible to determine the relative contributions of each of these pathways to total bile-acid biosynthesis, due to the nature of the data. Some values were obtained from patients whose gallbladders had been surgically removed; other patients would be atypical due to illness, and many data were obtained from experimental animals, which may metabolise these compounds differently from humans. Also, the exact order of many of the reactions is not known, since the intermediates may act as substrates for more than one enzyme. Further details for these reactions can be found in reviews by Chiang,[1] Moore *et al.*,[2] and Fuchs *et al.*[3]

1.3   Bile-Acid Regulation

1.3.1   Bile-Acid Receptors (FXR)

The following is a brief overview of events in the area of BA synthesis, transport and regulation. More detailed descriptions are given in Chapter 2 and can be found in the review by Redinger.[4] Bile acids from the enterohepatic circulation, upon returning to the liver, inhibit further BA synthesis by suppressing the rate-limiting enzyme CYP7A1 (cholesterol 7 alpha-hydroxylase) in the hepatocytes. They do this by binding to, and activating, FXR, the Farnesoid X receptor (NR1H4), a bile-acid receptor expressed in the liver, gut, kidneys and adrenal glands. In the liver, when bound to BA, FXR acts as a transcription factor to cause a feedback repression of BA synthesis.

FXR is one of 48 nuclear transcription factors so far identified, all of which reside inside the nucleus. They are involved in many biological processes, including cell growth and differentiation, embryonic development, and metabolism. They bind to ligands like bile acids, steroids and retinoids. BAs enter the cell and bind to FXR, resulting in conformational changes in structure, which allow them to react with, and influence, specific target genes. These reactions lead to the synthesis of inhibitory proteins, which repress the activity of the gene CYP7A1 (cholesterol 7 alpha-hydroxylase), the rate-limiting enzyme in BA biosynthesis. Another nuclear transcription factor to be discussed here is liver X receptor, (LXR), (NR1H3), which binds to cholesterol or its metabolic products. Other nuclear receptors like SHPs (small heterodimer partners), (NROB2), and SREBP, (sterol response element-binding proteins), have no identified ligands and are called orphan receptors.

FXR exists in alpha and beta forms and, when BA levels are high, the enzyme CYP7A1 is strongly repressed by a nuclear receptor cascade, in which activated FXR-alpha-BA ligand induces expression of the orphan receptor SHP. This shuts down the activity of another orphan receptor, liver receptor homologue-1, which is needed for CYP7A1 promoter activity.[5] Other indirect pathways also exist to repress CYP7A1 expression.[6] Hence, bile-acid levels in the liver control, through FXR, further bile-acid synthesis from cholesterol. All this ensures a constant, level of bile-acid production. Failure to achieve this regulated condition can lead to life-threatening conditions, such as liver, coronary, and cerebrovascular diseases.

Both FXR-alpha and -beta have been cloned. FXR has also been crystallised and its structure determined by X-ray crystallography. The shape of the cavity that holds a conjugated BA was determined.[7] FXR-alpha functions as a receptor for a wide range of bile acids, including cholic and deoxycholic acids, and their glycine and taurine conjugates. To keep BA levels constant in the liver, FXR can also induce the bile-salt export pump (BSEP) in the canalicular membrane. This is an ATP-dependent system and the main exporter of BA in the liver. In short, FXR suppresses the synthesis of new BAs and stimulates their billiary excretion, thus regulating BA levels and preventing excessive BA induced toxicity.

1.3.2 Cholesterol Receptors (LXR) (NR1H3) and (NR1H2)

Another receptor, LXR (Liver X receptor), also exists in alpha and beta forms, and acts as a receptor for cholesterol and its degradation products, which accumulate when cholesterol levels are high. LXRs are expressed in the liver and lower digestive tract, where they regulate cholesterol and bile-acid homeostasis. LXR-beta activates reverse cholesterol transport from the periphery to the liver.[8] LXR-alpha, which is found in the liver, promotes catabolism in the liver and drives catabolism of cholesterol to BAs. Its activation in the liver increases

cholesterol efflux and triglyceride production by inducing the expression of
SREBP-1c, as well as its target genes in the liver. LXRs also regulates fatty-acid
metabolism and exert anti-atherogenic effects by stimulating reverse cholesterol
transport and cholesterol excretion (*via* BAs).

LXR-alpha and -beta form heterodimers with the retinoid X receptor
(RXR), and are activated by cis-retinoic acid. The resulting compounds, RXR-
LXR-alpha and -beta heterodimers, interact with DNA response elements.
They bind to a D-4 element consisting of 2 hexanucleotides, direct repeat motifs
separated by 4 nucleotides (DR-4). These heterodimers are permissive, and can
be activated by ligands for both LXR and RXR.[9] The mechanisms whereby
these receptors interface with DNA are still being deciphered, but they appear
to be able to switch on CYP7A1 and drive BA synthesis. FXR and LXR,
working together, coordinately regulate BA synthesis and oxysterol homeostasis, as well as fatty-acid and triglyceride control. These factors are targets for
the development of therapeutic agents.

Entry of bile acids into the enterohepatic circulation from the gut is also
controlled by bile itself. BA absorption into the ileal epithelium depends on a
plasma membrane protein called the ileal bile-acid transporter (IBAT) gene
(SLC10A2). The promoter for this gene also binds the FXR-bile-acid complex
that starts the transcription that leads to synthesis of more transporters. As
FXR activity is stimulated by BA, there is a positive feedback from BAs to
IBAT, leading to up-regulated BA absorption and transport. Thus, we see bile
acids have the potential to control their own reabsorption *via* a protein feedback mechanism (FXR-IBAT) as well as controlling their own catabolism in
both a positive manner (LXR) and a negative manner (FXR-CYP7A1 and
FXR-BSEP). The processes involved are obviously far more complicated than
described here and are further explored in Chapter 2.

1.4   Chemistry of Bile Acids and Their Effects on Digestion

BA molecules are wedge-shaped, amphipathic structures, with a hydrophobic
side (represented by the steroid side of the molecule), and a hydrophilic side
(represented by the hydroxyl group, the amide carbonyl, and the ionised acidic
groups of either glycine or taurine).[10] The hydrophobicity of the bile acids may
well be linked to their intrinsic toxicity, with the more hydrophobic BAs being
more toxic. The hydrophobicity is inversely related to the number of OH groups.
Therefore LCA with only one OH group is highly hydrophobic and highly toxic,
whereas, DCA and CDCA with 2 OH groups and CA with 3 OH groups are
decreasingly hydrophobic and decreasingly toxic. The relative toxicity and
bioreactivity of the different BAs are discussed in detail in later chapters.

Four to eight hundred mls of bile are secreted daily in humans and the
contained BAs are strong detergents. They can be cytotoxic to the mucosal cell
membranes, and can adversely affect many tissues, both intra- and extra-
cellularly. Therefore, many strategies have evolved to control their distribution,

and maintain their concentration within narrow limits, to avoid cellular injury. For instance, bile is released into the small intestine only when there is food present (*via* cholecystokinin stimulated gallbladder contraction). The conjugated BAs are secreted into the duodenum as bile-acid anions, which mix with ingested food as it passes by. These BAs are conjugated and are thus largely impermeable to the cell membranes; hence, conjugated bile acids cannot leave the lumen of the upper GI tract. (The colon may be excluded from this since its bacterial flora can efficiently deconjugate them.) BAs are also signalling molecules, as described earlier,[1] that activate several nuclear receptors, and regulate many physiological pathways and processes to attain BA synthesis and cholesterol homeostasis. BAs also can induce signalling effects indirectly via their biological effects within the cell (*e.g.* the generation of ROS). These mechanisms, important in cancer development, are discussed in great detail in later chapters.

Bile acids help in the digestion of lipids and the products of digestion include dietary cholesterol, phospholipids, bilayers and fatty acids coming from the enzymatic breakdown of triglycerides. Association of these lipid derivatives with BAs forms mixed micelles, involving up to 40 BA molecules. The micellar mixture continues down the GI tract to the jejunum, where the contained lipids may diffuse into the epithelium to the portal veins. The micelles continue down to the distal ileum, where about 95% of the BAs are reabsorbed, and sent to the liver via the portal vein. This occurs several times during a typical high-fat meal, and forms the enterohepatic circulation (from the gallbladder, to the ileum, to the portal vein, and back to the liver). This cycling conserves BAs, thus avoiding the need to synthesise new BAs for each meal. The remaining $\sim 5\%$ of bile/micelles enter the colon, where the colonic bacteria break them down to lithocholic acid and deoxycholic acids, which are excreted in faeces. In humans, the faecal BAs are all deconjugated, due to an efficient bacterial enzyme system that deconjugates them and removes the 7-hydroxyls from the molecule. This represents the only time cholesterol (in the form of deconjugated faecal bile acids) is excreted from the body.

Deconjugation and dehydroxylation reactions occur in the colon, leading to the formation of dozens of new distinct BAs, by the action of the colonic bacteria. The final products enter the enterohepatic circulation and reach the liver where they are reconjugated mostly to either glycine or taurine. Some lithocholic acid, the most toxic substance produced in the body and a known carcinogen, enters the liver where it is sulfated or esterified to glucuronic acid and excreted.

1.5   Micelles

BAs in aqueous solution spontaneously aggregate to form micelles, these also mix with lipid products during digestion to form mixed micelles and enhance absorption. Their general shape is cylindrical, and can become worm-like, depending on the lipid-to-bile ratio. A micelle is pictured in Figure 1.2. Hjelm *et al.*[11] describe micelles as; "having the polar lipids arranged radially, with their hydrophilic heads facing outwards into the aqueous phase. The BA molecules are

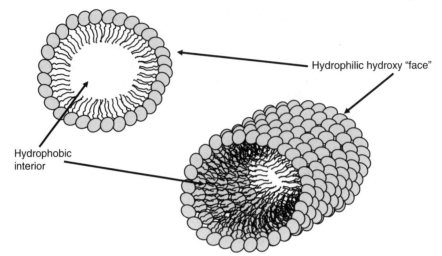

Hydrophilic hydroxy "face"

Hydrophobic
interior

Figure 1.2   Structure of a mixed bile-acid/fatty-acid micelle, whereby the hydrophilic
(OH groups of BA) are radially arranged on the outside of the micelle and
the hydrophobic moieties are arranged on the interior. As well as a classic
micelle, a cylindrical mixed micelle structure is also shown.

arranged perpendicularly between the polar heads. The hydrophobic faces of the
BA molecules rest like a wedge between the heads of the alkyl chains of the lipid
molecules; the hydrophilic face of the BA molecule faces the aqueous environment." This structure is the same for all micelles and they all have a negative
surface charge. Micelles also serve to transport lipids and vitamins in the GI tract.

Micelles tend to aggregate, and there are many ways to measure their
concentration, including surface tension measurements.[12] The midpoint of the
concentration range over which micellar aggregation occurs is called the critical
micellar concentration (CMC). Below the CMC, added bile-salt molecules
dissolve in the form of monomers; above the CMC, added bile-salt molecules
form micelles, leaving the monomeric concentration essentially constant. The
pH at which CMC formation occurs is called the critical micellar pH, (CMpH).
Table 1.1 lists values for some of the bile acids mentioned in this review.

Another term frequently used in this discussion is the pKa. Its relationship to
pH is described in the Henderson–Hasselbalch equation. Some modifications to
this equation have been made[12] to allow the calculation of many other physical/
chemical values, including the CMpH (Table 1.1).

1.6   Biochemical Properties of Bile Acids and Their Effects on the GI Tract

Table 1.1 lists some of the characteristics of the more common bile acids, which
are divided into 3 main classes: free bile acids, glycine and taurine conjugates.

1.6.1 Free Bile Acids

These include cholic and chenodeoxycholic (primary), and deoxycholic and lithocholic acids (secondary). Their pKa values range from 5.2–6.2 (Table 1.1) and they account for ~2% of bile. They are present in the enterohepatic circulation and are precipitated at low pH. Free BAs (like DCA) are formed in the lower GI tract and are usually absent from the upper GI tract, but in patients with less acidic stomach environments (through acid-suppression medication or through loss of acid-secreting glands), the stomach pH can reach almost neutral pH values, allowing gastric bacteria to proliferate, and they can deconjugate any conjugated BA reaching the stomach from the duodenum.[16] Hence, free bile acids can be present in both the upper and lower GI tract, but the efficient liver-based conjugation process is constantly converting them to their conjugated counterparts.

1.6.2 Glycine-Conjugated BAs

With pKas of 3.8 to 4.8, these are the most abundant conjugated bile acids, (representing >70% of bile). It should be noted that conjugation restricts their entry into the epithelial cells, ensuring that they remain within the intestinal lumen and do not leak into the intra-cellular spaces to damage other organs. However, at pH values approaching their pKa values, they become un-ionised and can cross intestinal membranes to a certain extent.

1.6.3 Taurine Conjugated dBAs

Representing >20% of bile, are strong sulphonic acids with detergent properties. They are soluble in the normal acidic stomach, with pKa values of <2.[12] Thus they can partially enter the gastric epithelium. It is also known that epithelial diffusion barriers can be broken by BAs,[17] which further allow them entry into the epithelium.

1.7 The Effect of pH on Bile-Acid Solubility

Free and glycine-conjugated BAs are only slightly soluble in acid solutions. As the pH is increased, the solubility will increase.[12] This is a very important characteristic since it describes the solubility characteristics of the major BAs, and, it also explains their potential to enter the epithelium at physiological pH ranges.

Occasionally, free BAs and glycine-conjugated BAs are found in the stomachs of normal volunteers.[16] Normal stomach acidity will precipitate them, whereupon they will leave the stomach along with the rest of the partially digested food. If not precipitated by stomach acidity, these BAs can also enter the oesophagus of patients who suffer from the reflux disease GORD (gastro-oesophageal reflux disease). Approximately 87% of these GORD patients were found to have BAs in their oesophagus,[18] mostly glycine conjugates.[19] Acid-suppressant therapy, the only medication used by these patients, will keep the pH of the refluxate at pH > 5 for over 20 h[20] and this may exacerbate GORD by preventing bile-acid precipitation in the stomach. The role of BAs in oesophageal cancer is described in some detail in Chapter 6.

From the above, it can be seen that different types of bile acids (free bile acids, taurine and glycine conjugates) will have access to the epithelial cells along the GI tract at normal physiological pH. For example, in the acidic stomach taurine conjugates would be soluble and potentially membrane permeable. In the more neutral pH small and large bowel, the free bile acids and the glycine conjugates would be soluble and permeable. Therefore, these different bile acids have the potential to start carcinogenesis across the whole GI tract, the bile type responsible being determined in each case by their solubility characteristic, their conjugation status and by their bioreactivity.

1.8 Potential Therapies for the Deleterious Effects of Bile Acids

Makeshima and his associates[21] postulated that LCA, a known colon carcinogen, is structurally similar to vitamin D, and like Vitamin D, it can activate the Vitamin D receptor, VDR. This would activate the gene CYP3A to make Cytochrome 450 enzymes to detoxify the LCA. Thus, adequate amounts of Vitamin D in the diet would protect against LCA-induced cancer, with the caveat that too much Vitamin D would have a potentially toxic hyper-calcemic effect. Experiments to find a drug to detoxify LCA, without at the same time affecting the calcium response are underway. This highlights the need for physiological balance. Bile acids play a fundamental role in normal human metabolism, excess BAs can be harmful, but so can the reduction of BA concentration, as is evidenced by the range of malabsorption diseases induced by lack of bile-acid absorption in the ileum.

The effect of antibiotic treatments on BA levels and the downstream effects of BAs are unknown. As bacterial de-conjugation and dehydroxylation is central to establishing and maintaining the normal bile pool, their decimation after antibiotic treatment can cause severe disruption. On the one hand, the reduction in free bile-acid production in the germ-free colon of patients on antibiotics could reduce the risks of carcinogenesis of the GI tract, through the action of free bile acids like DCA. It has certainly been shown in animal models that antibiotics increase the levels of conjugated bile acids in the lower GI tract.[22] However, the GI tract is not used to dealing with conjugated bile acids either and an increase in the level of conjugated bile in the lower GI tract could promote carcinogenesis at this site and indeed increased levels of conjugated BAs in the serum may drive carcinogenesis at other sites.[23]

Excessive amounts of BAs can accumulate in the GI tract (*e.g.* as a result of gallbladder surgery or cholecystectomy). These can be treated by the use of polymeric compounds which serve as ion exchangers, exchanging anions (such as chloride) for BAs. These compounds are known as BA sequestrants, and they absorb BAs from the enterohepatic circulation, whereupon they can be excreted with the faeces. Problems associated with these medications include diarrhoea, flatulence, cramps, *etc.* Again highlighting the balance needed in maintaining physiological levels of BAs. Further dietary modulation of BA levels is being investigated. Fibre content of diets may potentially reduce overall BA levels by binding to and promoting the excretion of BA. However, there is controversy in this area and further research is needed. The role of probiotics in altering the bile-acid pool and specifically the DCA to CA ratio is another area of research that may yield interesting results in the near future.

Statins are compounds that inhibit HMG-CoA reductase, the rate-limiting enzyme in cholesterol biosynthesis, and they are the world's best-selling drugs and are used for lowering cholesterol. Statins are well studied and are believed to be quite safe. Because they reduce the levels of cholesterol, the precursor of the bile acids, statins may be the ideal drugs to use for BA-lowering in these GI tract diseases.

1.9 Summary

Bile helps in the digestion and absorption of fats. Its constituent bile acids (BAs) have detergent properties, and some can be carcinogenic. BAs can act as signalling molecules, entering the nuclei and reacting with the nuclear receptors and this could enhance or reduce BA synthesis. In this way, they control their own levels as well as those of their precursor, cholesterol. This controls cholesterol homeostasis and BA and lipid synthesis.

Taurine-conjugated BAs ($>20\%$ of bile), are strong sulfonic acids, and are completely soluble in the normal stomach, while glycine-conjugated BAs ($>70\%$ of bile), are only slightly soluble at acid pH. As the pH increases, glycine-conjugated BA solubility increases, as does that of the free bile acids. Thus, when in high-pH solutions, these BAs are able to enter the epithelial cells of the GI tract and promote carcinogenesis.

Experimental evidence has highlighted a role of BAs in the induction and proliferation of Barrett's oesophagus, the induction of gastric epithelial damage, potentially inducing colorectal carcinogenesis and in numerous diseases of the gallbladder and liver. Keeping bile acids at physiological levels and preventing their build-up would overcome many of these problems. However, radical efforts to remove BAs, or completely alter the natural balance of subtypes present, may have serious side effects.

References

1. J. Chiang, Bile acid regulation of gene expression, Roles of nuclear hormone receptors, *Endocrine Reviews*, 2002, **23**(4), 443–463.
2. D. Moore, S. Shigeaki and D. Wen Xie, The NR1H and NRR1I Receptors: Constitutive Androstane Receptor, Pregnene X Receptor, Farnesoid

X Receptor alpha and beta, Liver X Receptor alpha and beta, and Vitamin D Receptor, International Union of Pharmacology, LXII.

3. M. Fuchs, Bile Regulation of Hepatic Physiology 3, Regulation of Bile-acid synthesis: Past Progress and Future Challenges, *Am. J. Physiol. Gastrointest. Liver Physiol.*, 2003, **284**, 551–557.

4. R. N. Redinger, The coming of age of our understanding of the enterohepatic circulation of bile salts, *The Am. J. Surg.*, 2003, **185**, 168–172.

5. A. Shulman and D. Mangelsdorf, Retinoid X Receptor Heterodymers in the Metabolic Syndrome, *N. Engl. J. Med.*, 2005, **353**, 605–615.

6. E. Scotti, F. Gilardo, C. Godio, E. Gers, J. Krneta, N. Mitro, E. De Fabiano, D. Caruso and M. Crestani, Bile acids and their signalling pathways: eclectic regulators of diverse cellular functions, *Cell Mol. Life Sci.*, 2007, **64**, 2477–2491.

7. L.-Z. Mi, S. Devarakionda and J. M. Harp, Structural basis for bile-acid binding and activation of the nuclear receptor FXR, *Mol. Cell*, 2003, **11**, 1093–100.

8. P. Tontonoz and D. J. Mangelsdorf, Liver Receptor Pathway in Cardiovascular Disease, *Mol. Endocrinol.*, 2003, **17**, 985–993.

9. B. A. Janowsky, P. J. Wiley and T. R. Devi, An Oxystyerol-Signalling Pathway, Mediated by the Nuclear Factor LXR-alpha, *Nature*, 1996, **383**, 728–31.

10. A. Hofmann, Bile acids, The Good, the Bad, the Ugly, *News Phys. Sci.*, 1999, **14**(1), 24–29.

11. R. Hjelm, C. D. Schteingart and A. Hofmann, Form and Structure of Self-assembling Particles in Monoolein-bile salt Mixtures, *J. Phys. Chem.*, 1995, **99**, 16395–400.

12. A. Hofmann and K. J. Mysels, Review: Bile Acid Solubility and Precipitation in vitro and in vivo: The Role of Conjugation, pH, and Ca ions, *J. Lipid Res.*, 1992, **33**, 617–626.

13. A. Roda, A. F. Hofmann and K. J. Mycels, The influence of bile salts structure on self-association in aqueous solution, *J. Biol. Chem.*, 1983, **258**(10), 6370–83.

14. P. P. Nair and D. Kritchevski, *The Bile Acids' Chemistry, Physiology and Metabolism: Volume 1, Chemistry*, Plenum Press, NY, 1971, p. 289.

15. A. Hofmann and D. M. Small, Detergent properties of Bile Salts: Correlation with Physiological Functions, *Annu. Rev. Med.*, 1984, **18**, 333–376.

16. J. Theisen, D. Nehra and D. Citron, Suppression of Gastric Acid Secretions in Patients with Gastroesophageal Reflux Diseases Results in Gastric Bacterial Overgrowths and Deconjugation of Bile Acids, *J. Gasterointest. Surg.*, 2000, **4**(1), 50–54.

17. S. Batzri, J. Harmon, E. J. Schweitzer and R. Toles, Bile-acid accumulation in the Gastric Mucosal Cells, *Proc. Soc. Exp. Biol. Med.*, 1991, **197**(4), 393–9.

18. D. Gotley, A. P. Morgan and M. J. Cooper, Bile-acid concentration in the Refluxate of Patients with Reflux Esophagitis, *Br. J. Surg.*, 1988, **75**, 587–590.

19. W. K. Kauer, J. H. Peters, T. R. DeMeester and A. P. Ireland, Composition and Concentration of Bile Acid Reflux into the Esophagus of Patients with Gastroesophageal Reflux Diseases, *Surgery*, 1997, **122**(5), 874–81.

20. Jai Moo Shin and G. Sachs, Restoration of Acid Secretion Following Treatment with Proton Pump Inhibitors, *Gastroenterology*, 2002, **123**, 1588–97.

21. M. Makeshima, T. T. Lu and W. Xie, Vitamin D as an Intestinal Bile-acid sensor, *Science*, 2002, **296**(5571), 1313–6.

22. J. Guban, D. R. Korver, G. E. Allison and G. W. Tannock, Relationship of dietary antimicrobial drug administration with broiler performance, decreased population levels of Lactobacillus salivarius and reduced bile salt deconjugation in the ileum of broiler chickens, *Poultry Science*, 2006, **85**, 2186–2194.

23. D. Stamp, Antibiotic therapy may induce cancers in the colon and breasts through a mechanism involving bile acids and colonic bacteria, *Medical Hypotheses*, 2004, **63**, 555–556.

CHAPTER 2
Bile-Acid Physiology and Measurement

PETER E. ROSS

Division of Pathology and Neurosciences, University of Dundee Ninewells Hospital & Medical School, Dundee, DD1 9SY, Scotland

2.1 Bile-Acid Physiology: The Enterohepatic Circulation

For around 1500 years circulation of blood and bile was considered in terms of the descriptions and thoughts of Aristotle, Hippocrates and Galen until the 17th century, a period when anatomists made great strides. William Harvey described the circulation of blood while Diemerbroek wrote that "bile enters the gallbladder to acquire greater strength and digestive power". At this time it was believed that bile was simply excreted, as indicated by Francis Glisson who wrote of bile "solely an excrementitious fluid". This view was dismissed by van Reverhorst who used the cannulated bile duct of a live dog to demonstrate that bile secretion into the duodenum would exceed the bile excreted from the gut. This idea was developed by Borelli who was able to estimate a biliary re-circulation rate of some 16 or 17 times. This is known to be high, but his thoughts 400 years ago on the role of bile in fat digestion were astute, including the belief that bile mixed with digested food in the duodenum and passed through pores to enter the mesenteric veins and subsequently the portal blood for return to the liver.

In the mid-18th century Edward Barry further developed thoughts on bile circulation and predicted a role for bile acids 80 years before von Liebig introduced the generic name bile acids. Some 100 years later absorption of bile

Issues in Toxicology
Bile Acids: Toxicology and Bioactivity
Edited by Gareth Jenkins and Laura J. Hardie
© Royal Society of Chemistry, 2008

acids by the small intestine was demonstrated in isolated bowel segments before Schiff elegantly demonstrated the recirculation by returning ox bile to the portal blood of guinea pigs and showing ox bile acids in the guinea pig bile subsequently secreted. A more detailed treatise of this historical summary can be found in the review by Reuben.[1]

Bile secretion has a central role in liver physiology, of particular importance for excretion of both endogenous and exogenous compounds, and in the absorption of lipid from the gut. Bile acids, the major organic components of bile, are conjugated to glycine or taurine, which reduces their ionisation constants so that they are ionised at physiological pH and present as acids. For the purpose of this chapter they will be referred to generically as bile acids and in the same way the abbreviation for human proteins will be used irrespective of species. The amphipathic properties of bile acids have been well described and their vital role in the absorption of dietary fats is unequivocal, but the ability to solubilise fats into emulsions for absorption is also potentially catastrophic for cells that depend on lipid membranes for their integrity. Consequently, the process from synthesis through the enterohepatic circulation to excretion must be tightly regulated using a complex arrangement of rapid chemical pumps and slower mechanical pumps, the gallbladder and small bowel. The hepatocyte, the only site of synthesis of bile acids, is responsible for the maintenance of a steady-state pool size by *de novo* synthesis from cholesterol to replace up to 0.6 g lost per day by faecal excretion. In light of this enterohepatic circulation it is clear that bile-acid content of the hepatocyte reflects contributions from bile acids taken up from portal blood and from *de novo* synthesis.

2.2 Transport in Blood

In plasma, bile acids circulate mainly bound to albumin although Kramer *et al.*[2] used photoaffinity reactions with radiolabelled taurocholic acid to label proteins in fasting serum. This approach confirmed binding by albumin but also demonstrated binding to high-density lipoprotein (HDL). Most taurocholic acid in HDL was associated with the lipid component, around 15% associated with the protein. Roda and co-workers[3] used radiolabelled bile acids in equilibrium dialysis studies to demonstrate that increasing hydrophobicity was associated with higher affinity constants and unconjugated bile acids showed a slightly higher affinity than conjugated molecules. This binding minimises free concentrations of bile acids, particularly the more toxic compounds, but also illustrates the efficiency of uptake by the hepatocyte as concentrations in peripheral plasma are low.

2.3 The Hepatocyte

Protein-bound bile acids in plasma are removed with high efficiency by the hepatocyte, partly due to the liver sinusoids that allow protein-bound material to enter the space of Disse and thus expose these compounds to the basolateral membrane of the hepatocyte. Efflux of bile acids from the basolateral membrane is negligible under normal physiology but may occur in cholestasis, which will be discussed briefly later. Early work by Schwarz *et al.* using isolated rat liver cells showed a 200-fold concentration of physiological levels of taurocholate by a process competitively inhibited by taurochenodeoxycholate. Ouabain, an inhibitor of the basolateral Na^+/K^+ ATPase that creates the Na^+ gradient, inhibited the uptake by around 75%, as did replacement of extra-cellular Na^+ by either K^+ or sucrose. From this the authors concluded that the uptake of bile acids is carrier mediated and involved the co-transport of Na^+ ions.[4] The process is efficient, with over 80% extraction of bile acids in one pass as reported by Reichen and Paumgartner who also showed the co-transport of Na^+ ions.[5]

2.4   Sodium Taurocholate Co-transporting Protein (NTCP)

Photoaffinity reactions were used to attach radiolabelled bile acids in an effort to identify transporter proteins at the sinusoidal plasma membrane. Two proteins were identified, of 68 and 54 kDa. The 68-kDa protein was water soluble and identified as albumin, presumably associated with the membrane, while the 54-kDa protein could only be solubilised in Triton X100 and is an integral membrane protein.[6] This transporter, now known as the Na^+-taurocholate co-transporting protein (NTCP)[7] is driven by the sodium gradient across the hepatocyte membrane and concentrates bile acids within the hepatocyte. NTCP transports both conjugated and unconjugated bile acids and is a transporter primarily for bile acids although oestrone 3-sulphate, for example, is also transported.[8] As shown by the studies of Schwarz's group and Reichen and Paumgartner[4,5] around 75% of taurocholate uptake by the hepatocyte is Na^+ dependent. Much of the work defining the role of NTCP as the prime transporter for bile acids has been circumstantial, including similar Km values for bile-acid transport in hepatocytes and NTCP transfected cells,[9] parallel reductions in expression of the transporter and taurocholate transport in cultured rat hepatocytes[10] and similar developmental profiles of the transporter and taurocholate transport.[11] The use of anti-sense oligonucleotides reduced taurocholate transport in transfected *Xenopus* oocytes[12] appeared to have provided the final proof but Wolkoff and Cohen cast doubt on this by pointing out that the anti-sense oligonucleotide used was not specific for NTCP but also interfered with expression of several rat proteins along with *Xenopus* oocyte proteins involved in signalling pathways.[13]

Although absolute proof is still lacking it seems clear that NTCP is the major sodium-dependent transporter of bile acids, although a minor role for other proteins cannot be excluded. It has now been isolated from rat,[14] mouse,[15] rabbit[16] and human.[17] The rat polypeptide was first expressed in *Xenopus laevis* oocytes and shown to be a 362 amino acid glycoprotein with 7 or 9 transmembrane domains[18] located at the basolateral, or sinusoidal, membrane with a cytoplasmic C-terminus and extra-cellular glycosylated N-terminal.[19,20] Peptide antibodies showed that the transporter is only expressed at the sinusoidal membrane of the hepatocyte with no expression in ileal tissue.[19] There is a report by von Dippe and colleagues[21] describing stable transfection of microsomal epoxide hydrolase (MEH) that facilitated Na$^+$-dependent transport of bile acids in MDCK cells in the absence of NTCP. However, the physiological significance remains unclear as this enzyme is normally membrane bound in the endoplasmic reticulum and may be involved in transport of bile acids and related molecules across the endoplasmic reticulum membranes, perhaps in the reconjugation of unconjugated bile acids returned by the enterohepatic circulation. Alves and colleagues found different topological orientation in right side out smooth endoplasmic reticulum membrane vesicles[22] that could indicate a role for vesicular uptake. Care must be exercised in interpretation of transport data from transfection models as transport was inhibited by albumin, in contrast to NTCP. The presence or absence of albumin is also important in *Xenopus* oocytes transfections as NTCP can transport bromosulphthalein and oestrone 3-sulphate in the absence of albumin but not when albumin is present.[18] It is possible that other transporters function only in abnormal physiological states, such as cholestasis, with raised bile-acid concentrations, or over-expression of transporters in cell line models.

2.5   Organic-Anion-Transporting Peptide (OATP)

An unrelated transporter was subsequently identified that is responsible for the Na$^+$-independent transport of taurocholate[23] and forms part of the super-family of organic-anion-transporting polypeptides (OATP).[24] These are independent of sodium and show a wide specificity, transporting unconjugated bile acids and other organic anions including physiological toxins and xenobiotics. These Na$^+$-independent processes are found in lower vertebrates that lack Na$^+$-dependent transporters, suggesting that these Na$^+$-dependent transporters evolved more recently.[24] The mechanism of transport is different from NTCP, being electroneutral as each bile acid crossing into the hepatocyte is balanced by efflux of a bicarbonate ion[25] or glutathione.[26] These uptake mechanisms are summarised in Figure 2.1.

2.6   Specificity of OATP and NTCP Transporters

The first studies of specificity were carried out using cholate, the glycine and taurine conjugates and taurine conjugates of the dihydroxy bile acids chenodeoxycholate and ursodeoxycholate.[8,18] Kramer and colleagues prepared plasma membrane vesicles from rat liver and compared bile-acid transport with values from CHO cells stably expressing NTCP.[27] This work established that transport by the liver enzyme was maximal when 2 hydroxyls were present,

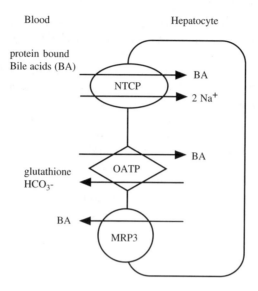

Figure 2.1 Hepatocyte basolateral bile acid transporters. Protein-bound bile acids returning in portal blood are taken up by the hepatocyte *via* the sodium taurocholate co-transporting polypeptide (NTCP) and organic-anion-transporting polypeptide (OATP). In cholestasis bile acids may be returned to blood by the multi-drug-resistance-associated protein 3 (MRP3).

either 3, 7 or 12. While the 3α hydroxyl, present in all bile acids, would be expected to exert a major effect on transporter affinity this was not the case. Removal of this hydroxyl moderately reduced transport, as did removal of taurine to form unconjugated cholate. Interestingly, NTCP showed the highest affinity for bile acids with a 6 hydroxyl although this is uncommon in man. A detailed comparison of specificities was also carried out using HeLa cells stably transfected to express rat NTCP or rat OATP where expression was under the regulation of a zinc-inducible promoter.[28] OATP and NTCP-transported dihydroxy bile acids but NTCP is more efficient with trihydroxy bile acids than OATP. It is clear, however, that both transporters have a wide specificity for bile acids, steroids and thyroid hormone[29] with OATP showing higher affinities for conjugated steroids such as oestrone-3-sulphate, oestradiol-17β-glucuronide and dehydroepiandrosterone sulphate.[30]

Both specificity studies confirmed that bromosulphthalein (BSP) competitively inhibited taurocholate transport by NTCP and OATP. This is in conflict with reports that BSP transport was not sodium dependent, suggesting that OATP was responsible.[31,32] The reason for this difference is not clear but may reflect differences in the approaches, using isolated rat hepatocytes or transfection to produce cells that stably express the protein. Choice of cell line may also be important as expression of MEH also showed differences, with no demonstrable Na^+-dependent transport of taurocholate in Syrian hamster kidney cells or oocytes[33] but Na^+-dependent transport was shown in Mardin–Darby canine kidney (MDCK) cells.[34] Caution must therefore be exercised when physiological transport parameters are related to transport characteristics determined with stably expressing cells. Albumin may also play a role in this difference as discussed earlier.[13] It is also important to note that Hata and colleagues compared cells transfected with rat and rabbit NTCP and found marked differences in uptake, showing that choice of species may impact on specificities.[28]

Bile acids are normally removed with high efficiency from blood but under conditions of cholestasis, where transfer of bile acids across the hepatocyte and secretion into bile is impaired, there may be a secretion from the hepatocyte into blood. There is evidence that multi-drug-resistance-associated protein 3 (MRP3), a basolateral transporter with low expression in normal hepatocytes, can transport glycocholic acid but not the tauro conjugate.[35] MRP3 is up-regulated in obstructive cholestasis and may export bile acids from the hepatocyte across the basolateral membrane to blood,[36] limiting the concentration of bile acids within the hepatocyte. However, this may be species specific as mice lacking MRP3 expression experienced no change in urinary excretion of bile acids when bile-duct ligation was used to model cholestasis.[37]

2.7 Regulation of Expression

Function of this transporter is regulated by altered gene expression resulting in changed synthesis of the protein and changes in the insertion into the basolateral membrane. The promoter for NTCP directs expression only in hepatocytes although small amounts of expression are seen in Caco2 and MDCK cells; these cells are derived from epithelia of tissues capable of bile-acid transport and may therefore have factors that recognise elements in the rat NTCP promoter.[38] These authors identified one of the factors as the hepatic nuclear transcription factor HNF1 but further work has shown that this regulates expression in rats but not the human or mouse promoters of NTCP. Indeed, HNF1α, HNF4α and the retinoid X receptor/retinoic acid receptor dimer (RXRα/RARα) were shown to activate the rat NTCP promoter but had no effect on the human or mouse promoter.[39] Similarly, bile acids were shown to down-regulate expression of NTCP in rats but this was not found with either human or mouse NTCP.[39] This indicates that there are significant differences between rat, mouse and human NTCP promoter regions but does not preclude bile acids exerting some control over expression of NTCP as bile-acid signalling may indirectly effect nuclear factors that regulate expression of NTCP. At a post-transcriptional level cAMP is now known to have two effects. Firstly, there is a rapid insertion of NTCP into the basolateral membrane from the pool of vesicular NTCP and secondly cAMP appears to dephosphorylate serine 226 that increases insertion into the basolateral membrane leading to greater bile-acid transport. The phosphorylation did not seem to alter activity of the NTCP.[40]

OATP expression is controlled by similar nuclear factors to NTCP but, unlike NTCP, the activity is down-regulated by phosphorylation of serine following activation of the protein kinase C pathway indicating control of activity without removal from the membrane.[41]

2.8   Transport across the Hepatocyte

The identification and cloning of membrane-bound transporters has established how bile acids are removed from plasma and enter the hepatocyte but details are less clear on the intra-cellular transport of these compounds across the hepatocyte. It is unlikely that free bile acids will exist in the cytoplasm, due to their toxicity and membrane-damaging nature. Early work[42] using electron microscopic auto-radiography after administration of a bile-acid labelled tyrosine conjugate suggested an involvement of smooth endoplasmic reticulum and Golgi apparatus.[43] Simion and colleagues used hplc analysis of fractionated rat liver and found around 60% of bile acid was associated with the cytoplasm.[44] Taurocholate binding studies showed a high-affinity binding to Golgi and plasma membranes and the cell supernatant along with low-affinity binding in all membranes. These authors attributed the low-affinity binding to nonspecific portioning into membranes and the higher-affinity binding sites showing that these organelles play a role in transport of bile acids across the hepatocyte.

Vesicular transport of bile acids has not been demonstrated under normal conditions, shown by using isolated rat hepatocyte couplets and fluorescently labelled bile acids. In these experiments confocal microscopy found no evidence of sequestering into clusters and colchicine disruption of microtubular function did not affect bile-acid transport.[45] This makes it unlikely that vesicle transport plays a role and it is now believed that bile acids traverse the hepatocyte by diffusion through the cytosol while bound to soluble proteins. It is worth considering the caveat that fluorescently labelled bile acids, while very useful tools, do differ structurally from endogenous bile acids with increased hydrophobicity leading to greater retention by cells.[46]

Glutathione S transferases bind bile acids *in vitro* but doubt has been cast over whether this happens *in vivo* as these enzymes were not labelled by fluorescently labelled bile acids in experiments to identify the carrier proteins[47] but may play a role with the raised levels in cholestasis. Liver fatty-acid-binding protein has been shown to bind bile acids by using a displacement assay with fluorescent fatty-acid ligand. This work clearly showed displacement to be directly related to hydrophobicity, such that lithocholate conjugates had the greatest effect.[48] This may indicate a mechanism to minimise toxicity within the hepatocyte.

The enzyme 3α-hydroxysteroid dehydrogenase plays a key role in this transport across the hepatocyte. A particularly elegant experiment demonstrated the role of the 3α-hydroxysteroid dehydrogenase, by using [^{3}H] at the 3β hydrogen to show cyclical oxidation-reduction of the 3α-hydroxyl with no accumulation of 3-keto bile acids.[49] Confirmation was obtained by use of indomethacin, an inhibitor of 3α-hydroxysteroid dehydrogenase, which decreased secretion of glycocholate from rat hepatocytes.[50] Other dehydrogenases with mass similar to 3α-hydroxysteroid dehydrogenase show high affinity for bile acids but do not have activity towards the 3α-hydroxyl[51,52] and have not been shown to have a physiological role.

However, hepatic portal blood returns both conjugated and unconjugated bile acids to the hepatocyte and during transit these unconjugated bile acids are reconjugated. Unlike *de novo* synthesis, where conjugation occurs in the peroxisomes,[53] this occurs in the endoplasmic reticulum where both bile acid-CoA ligase and the bile acyl CoA:amino acid N-acyltransferase are found. It is believed that protein-bound bile acids reach the canalicular membrane by diffusion along the concentration gradient that is maintained by the bile-acid export pump (BSEP).

2.9   Bile-Acid Export from the Hepatocyte

In 1991 bile-acid secretion was shown to be energy driven by a 110-kDa glycoprotein that was dependent on ATP.[54] This protein was subsequently characterised as liver ecto-ATPase by Sippel and co-workers. However, while further work with COS cells showed that expression of ecto-ATPase enhanced secretion of bile acids[55] purified canalicular membranes lacking this enzyme efficiently exported bile acids showing that at least one other bile-acid transporter existed.[56]

2.9.1   Bile-Salt Export Pump (BSEP)

The gene for a new member of the P glycoproteins, a small family of transport proteins associated with multi-drug resistance, was isolated from mammalian liver and, lacking a function, named as sister of P-glycoprotein, a member of the ATP-binding cassette (ABC) transporter family (Abcb11).[57] Adapting techniques used to investigate the multi-drug-resistance genes Brown *et al.* used cholic acid methyl ester to load hepatoma-derived cells with cholic acid and to compare cells that became tolerant with naïve cells. This demonstrated that tolerance was linked to over-expression of ABC transport proteins in the 150–200 kDa range.[58] Some three years later reverse transcription-polymerase chain reaction on mRNA isolated from rat liver gave a fragment showing 88% homology with "sister of P glycoprotein" (spgp). In an elegant study Gerloff and colleagues isolated full-length spgp and demonstrated the ATP-dependent transport of bile acids in transfected Sf9 cells and in *Xenopus laevis* oocytes.[59]

When kinetics were compared for bile-acid transport for canalicular membranes and transporter expressed in Sf9 cells, rates were broadly similar, indicating that this transporter, now called bile-salt export protein (BSEP), is the major bile-acid exporter.[60] This is supported by the identification of mutations in this gene that cause a dramatic drop in secretion of bile acids into bile, less than 1% of the normal secretion rate.[61] However, the picture may be more complicated as mice with the BSEP gene knocked out cannot secrete tauro-cholate across the canalicular membrane, as predicted, but secrete muricholic acid and a tetra-hydroxylated bile acid into bile giving a concentration of bile acids around 30% of normal.[62] Muricholic acid ($3\alpha,6\beta,7\beta$) differs from cholic acid ($3\alpha,7\alpha,12\alpha$) in both position and orientation of hydroxyls and it seems plausible that a different transporter is responsible, but presence of the tetra-hydroxylated bile acid suggests it may be a changed pathway to limit intra-cellular bile-acid concentration and perhaps secretion of sulphated or glucur-onidated bile acids. This was recently supported by Lam *et al.* who investigated BSEP null mice and found markedly raised levels of multi-drug-resistance protein MDR1, structurally related to BSEP, but only moderate increases were found for MDR2.[63] See Figure 2.2 for a summary of apical transporters secreting into the canaliculus.

2.9.2   Regulation of BSEP

Transport of bile acids from blood into bile is rate limited by activity of the BSEP, and plays a crucial role both in control of intra-cellular concentrations

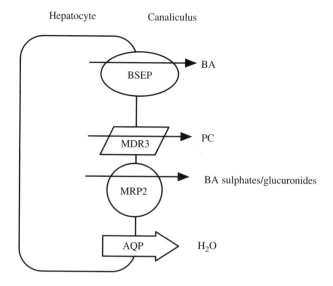

Figure 2.2   Secretion of bile acids and biliary components. Bile acids (BA) cross the hepatocyte bound to 3α-hydroxysteroid dehydrogenase and are exported into the canaliculus by the bile-salt export protein (BSEP). Phosphati-dylcholine (PC) from the inner leaflet of the apical membrane is flipped to the outer layer and interacts with bile acids secreted by BSEP. BA, PC, together with cholesterol from the membrane form mixed micelles that are not toxic to epithelial membranes of the biliary tree. Aquaporins (AQP) secrete water into bile.

of bile acids in the hepatocyte and in secretion of bile. Feeding of cholate to mice leads to an up-regulation of BSEP gene expression by the nuclear receptor FXR/BAR (Farnesyl X receptor/bile-acid receptor) which also neatly demonstrated the role of this nuclear receptor as a sensor for intra-cellular bile-acid concentration.[64] Regulation by FXR/BAR has also been shown in man,[65] confirming that bile acids returning to the liver control transcription of BSEP.

Post-translational changes provide a more rapid regulation of BSEP activity including insertion into or removal from the canalicular membrane. Insertion into the canalicular membrane is stimulated by intravenous infusion of cholate to rats or infusion into isolated perfused rat liver. By use of inhibitors such as wortmannin and colchicine the importance of phosphoinositide-3-kinase was demonstrated in cholate-induced insertion of BSEP into the canalicular membrane.[66] This also resolves the earlier confusion over trans-cellular trans-port of bile acids, with colchicine-reduced bile-acid secretion but fluorescent probes showed no association with the microtubules. The decrease in bile-acid secretion is due to microtubule disruption preventing insertion of ATP-dependent transporters, including BSEP. This suggests the presence of an intra-cellular pool of vesicular transporters that are available for rapid insertion into the canalicular membrane as required, a fact confirmed by use of canalicular membranes isolated from rat livers after treatment with taurocholate, dibutyryl cyclic AMP (dBcAMP, a soluble analogue of cAMP) and colchicine that showed increased transport function was linked to increased transporter pro-teins in the membrane. The increase, shown by Western blotting, was prevented by colchicine, indicating a role for microtubules in the process.[67]

Subsequent work by Misra and colleagues showed that administration of taurocholate led to activation of phosphatidylinositol-3-kinase (PI-3 kinase) activity and increased levels of ATP-dependent transporters in vesicles isolated from canalicular membranes. This treatment also raised levels of some proteins, including cell-adhesion molecule cCAM105, in vesicles isolated from both canalicular and sinusoidal membranes but the transporters were only found in canalicular membrane vesicles.[66] These authors confirmed the inhibitory action of colchicine, showing a decreased canalicular PI-3 kinase activity that indicates that this kinase is transported to the canalicular membrane by microtubules. As predicted, colchicine treatment leads to accumulation of ecto-enzymes, PI-3 kinase and cCAM105 at the basolateral membrane, supporting the belief that microtubule-based transcytosis transports these proteins from the basolateral membrane to the canalicular membrane. Wortmannin, a PI-3 kinase inhibitor, would have a multi-organ effect and therefore an isolated liver perfusion system was used to investigate the role of PI-3 kinase activity. When wortmannin was administered after taurocholate infusion, bile-acid secretion was reduced to 50% but there was no reduction in ATP-dependent transporters in canalicular membranes. However, if wortmannin was perfused before taurocholate then there was a marked reduction in these transporters, suggesting that bile-acid secretion is regulated by PI-3 kinase activity. A year later this group confirmed this finding when they showed that phosphatidylinositide 3,4 bisphosphate could ameliorate inhibition by specific PI-3 kinase inhibitors and therefore the lipid products of this kinase are active in control of bile-acid secretion.[68]

Cyclic AMP (cAMP) has been known to increase bile flow for some time, primarily in response to feeding, while dBcAMP enhanced hepatocyte secretion of bile acids by a mechanism involving sodium and calcium flux.[69] Wortmannin given before or after dBcAMP inhibited the up-regulation of bile-acid secretion, suggesting that the controlling influence is by the phosphoinositol pathway. However, recruitment of BSEP in response to cAMP was not inhibited by wortmannin, suggesting a different pathway from the PI-3 kinase pathway initiated by taurocholate.[70] This is in agreement with the report showing the effects of cAMP and PI-3 kinase to be additive.[67] The role of cAMP is complex due to the cross-talk between signalling pathways and those of PI-3 kinase products. PI-3 kinase is activated by cAMP, but it seems by a process that does not involve phosphorylation of tyrosines, at least in polarised cells.[70] Some care is required in interpretation of this data as cAMP activation of PI-3 kinase has been shown to be cell-type specific,[71] while the cAMP effect differs in cells over-expressing PI-3 kinase and Akt.[72]

There is evidence that cAMP stimulates incorporation of another transporter, the canalicular multi-specific organic-anion transporter (cMOAT) also called the multi-drug-resistance protein 2 (MRP2).[73] As with BSEP, cAMP up-regulates insertion into the canalicular membrane by a microtubular-based transcytosis. However, these authors suggest that the cAMP effect is not specific for insertion of MRP2 but is in fact a general stimulation of apically directed vesicles. This accounts for the increased apical levels of the canalicular Cl^-/HCO_3^- exchanger in response to cAMP[74] and enhanced biliary secretion of horseradish peroxide, a commonly used marker for micropinocytosis, which is compatible with greater exocytosis of vesicles at the apical membrane.[75]

Activation of protein kinase A by cAMP is known to stimulate sphingolipid transport to the apical membrane, a process that can occur either by direct transport from the Golgi to the apical membrane or by transcytosis from basolateral to apical membranes. Fluorescent sphingolipids showed that both pathways are increased by the dibutyryl form of cAMP and it seems reasonable that increased apical transporters occur by a similar mechanism.[76] Increased endocytosis at the apical membrane could provide a rapid means to reduce levels of transporters at the membrane yet sub-apical vesicles or an apical sorting compartment would provide a pool for rapid insertion under the control of cAMP or protein kinase A.[73] When oestradiol-17β-D-glucuronide was used to cause cholestasis in rats BSEP in the apical membrane was reduced by endocytosis, although this reduction was prevented by a cAMP analogue that would increase vesicular insertion of BSEP into the apical membrane,[77] suggesting that cholestasis is associated with down-regulation of BSEP. An earlier study using 17α ethinyloestradiol showed a similar effect with BSEP but also found a larger decrease in NTCP, an effect that would limit the concentration of bile acid within the hepatocyte, presumably to minimise damage.[78]

BSEP structure predicts multiple phosphorylation sites, including those for protein kinase C and A (PKC and PKA), within the predicted intra-cellular domains. However, PKA showed only weak phosphorylation of BSEP obtained by immunoprecipitation, indicating, at least *in vitro*, limited phosphorylation by PKA.[79] Bile acids and PKC are involved in hepatocyte apoptosis where PKCα, PKCδ and PKCε move to the plasma membrane in response to bile acids[80] and consequently PKCα and BSEP were co-expressed in Sf9 cells that led to an 18-fold increase in phosphorylation following activation of PKC with phorbol ester.[79] Treatment with tauroursodeoxycholic acid, a bile acid that increases bile flow, leads to PKCα moving to the plasma membrane,[81] while taurolithocholic acid reduces biliary secretion and is associated with movement of PKCε to the plasma membrane.[82] This suggests that bile acids initiate signalling by PKC to control BSEP function although the role of BSEP phosphorylation remains to be identified.

2.9.3 Other Transporters that Export Bile Acid from the Hepatocyte

A further ABC transporter, MRP2, is involved in apical transport of bile acids into the canaliculus as shown in Figure 2.2. This protein has a wide range of substrates, including bilirubin conjugates and glucuronide or sulphate conjugates of drugs.[83,84] MRP2 transports tauro or glycolithochic acid sulphated at the 3α-hydroxyl but cannot transport nonsulphated monovalent bile acids such as taurocholic acid.[85] This work established the role of 2 arginine residues, 586 and 1096 where the cationic charge blocks taurocholate transport. Mutation of these residues with neutral amino acids maintained transport of glutathione and glucuronide conjugates and facilitated transport of taurocholic acid. This is shown in Figure 2.2.

A related protein, MRP3, has similar structure to this mutated MRP2 and can transport taurocholic acid but mutation of the equivalent residue, leucine 1084, with lysine-blocked transport of taurocholic acid. In cholestasis there is an induction of MRP3 mRNA suggesting that this transporter is active, at least when bile-acid concentrations are raised within the hepatocyte.[86] This transporter function is shown in Figure 2.1.

2.10 Bile Secretion

Bile secretion is divided into two components, the bile-acid-dependent fraction (30–60%) and the bile-acid-independent fraction that is probably associated with biliary secretion of glutathione and inorganic solutes.[87] Considering the bile-acid-dependent component, secretion of bile acids into the bile canaliculi would be expected to cause damage to the canalicular membrane but this does not occur as the presence of bile acids in the canaliculi stimulates secretion of phospholipids and probably cholesterol. The presence of these molecules is protective against the toxic effects of bile acids[88] but was originally believed to reflect a detergent action on the canalicular membrane.[89] This detergent action mechanism does not explain why biliary phospholipids are predominately phosphatidylcholine (PC) (>95%) and in particular largely consist of molecules with a palmitoyl residue (C16) at the sn-1 position,[90] rather than reflecting the membrane composition of the canalicular membrane. Secretion of phospholipid into bile was attributed to phosphatidylcholine transfer protein that catalyses exchange of this phospholipid between intacellular membranes or out of the cell.[91]

The protein responsible for biliary secretion of phospholipids was subsequently identified following discovery of the ABC family member multi-drug-resistant 3 gene (confusingly MDR3 in man, Mdr2 in mouse) that had no known function, although it seems that this protein can also transport organic anions.[92,93] Creation of an Mdr2-null mouse showed that deficiency of this protein prevented secretion of phospholipids into bile and led to mild liver disease.[94] In man, deficiency leads to progressive familial intrahepatic cholestasis, a disease that will result in liver transplant.[95] The Mdr2-null mouse allowed investigation of the mechanism of action as lack of this transporter prevented the secretion of PC and caused a major reduction in cholesterol secretion, indicating that secretion of these two lipids is linked.[96] These authors also demonstrated the requirement for bile acids by cannulating the gallbladder to remove bile and prevent return of bile acids to the liver. The reduction in bile-acid pool size over time led to a similar time-dependent reduction in bile flow and both phospholipid and cholesterol content of bile. Infusing increasing levels of bile acid restored biliary levels of cholesterol and phospholipids in wild-type mice but not in the null mouse, indicating a requirement for both transfer protein and bile acids.

The phosphatidylcholine in bile is synthesised in the endoplasmic reticulum of the hepatocyte and must be transported to the canalicular membrane. One possibility involves the nonspecific phosphatidylcholine transfer protein but a mouse null for this protein did not show reduced phosphatidylcholine secretion into bile and there was no compensatory increase in other phospholipids transfer proteins.[97] However, the plasma membrane would receive a ready supply of phospholipid by insertion of vesicles, and the MDR3 protein translocates this molecule from the inner leaflet to the outer surface where there is contact with bile acids, as suggested by Smit and colleagues.[94] The role of this transporter is shown in Figure 2.2.

The presence of MDR3 to transfer PC from inner to outer leaflet might of itself explain the preponderance of PC in bile, as no transfer protein for sphingolipids has been identified. Sphingolipid transport to the apical membrane has been shown to occur both by direct and by transcytotic vesicular processes activated by cAMP.[76] Indeed, the content of sphingomyelin and cholesterol is higher in the apical membrane than the basolateral membrane with a concomitant reduction in membrane fluidity at the canalicular surface, and these regions of the membrane with sphingomyelin and cholesterol would also be resistant to micelle formation with bile acids.[98] Thus, the apical membrane may be envisaged as nonfluid regions of high sphingomyelin/cholesterol content and more fluid regions of PC/cholesterol. These PC/cholesterol regions interact with bile acids to form mixed micelles found in bile with MDR3 replenishing PC in the outer leaflet. However, there may be a component of micellar solubilisation by bile acids as livers from Mdr2-null mice secrete cholesterol but do not secrete PC when perfused with taurocholate.[96]

2.11 Water Secretion

Bile secretion involves co-ordinated activity of a number of membrane-located transporters and water movement into bile from the hepatocyte was believed to be simply a response to osmotic gradients involving vesicular traffic and diffusion across the lipid membrane.[99] More recently, it has been shown to involve the aquaporin (AQP) family of membrane proteins,[100] as shown in Figure 2.2. These proteins are responsible for transport of water across epithelial membranes in a variety of organs of which three were identified in hepatocytes. Of these, AQP0 and 8 were found intra-cellularly and at the apical membrane. Treatment with dibutyryl cAMP increased the levels of AQP8 at the canalicular membrane but had no effect on AQP0 and 9.[101] Cholestasis initiated by 17α-ethinyloestradiol is associated with a down-regulation of AQP8 expression, indicating that secretion of water is part of the dysfunction of this process.[102]

2.12 Cholangiocytes

Cholangiocytes are epithelial cells lining the bile ducts and possess transporters for a number of molecules allowing reabsorption of glucose, amino acids, glutathione and bile acids, in addition to absorption/secretion of fluid. Control of these functions is wide and varied, but do include effects by bile acids. In 1997 the role of the apical sodium-dependent bile-acid transporter (ASBT), already identified in the enterocyte, was identified in the apical membrane of cholangiocytes, indicating the ability to absorb bile acids from the bile-duct lumen (see Figure 2.3).[103] Uptake of bile acids by the ASBT activates a process called cholehepatic shunting that increases biliary lipid and fluid secretion by the hepatocyte.[104] In bile duct obstruction bile-acid uptake by ASBT stimulates proliferation by activation of a PI-3 kinase pathway.[105] This mechanism was extended to show that bile acids activate the epidermal growth factor receptor (EGFr) by a process involving transforming growth factor α (TGF α) and a matrix metalloproteinase to release the TGF α.[106] Bile-duct proliferation would lead to increased cycling of bile acids from cholangiocyte to hepatocyte and greater potential for conjugation to reduce toxicity. Bile acids are secreted from the basolateral membrane by the MRP3 transporter, although there is evidence that a truncated, alternatively spliced, ASBT (tASBT) can transport taurocholic acid across the basolateral membrane,[107] as shown in Figure 2.3. More recently, a heteromeric organic-solute transporter (OST) comprising α and β forms has been implicated in Na⁺-independent efflux of bile acids across the basolateral membrane.[108] This is considered further when discussing efflux from the enterocyte.

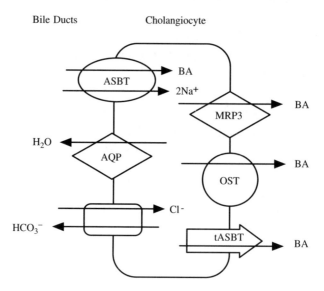

Figure 2.3 Absorption of bile acids by the cholangiocyte in the cholehepatic shunt. Bile acids are absorbed at the apical membrane of the cholangiocyte by the apical sodium-dependent bile-acid transporter (ASBT) that causes cholehepatic shunting of bile acids back to the hepatocyte. Absorbed bile acids are exported across the basolateral membrane by multi-drug-resistance-associated protein 3 (MRP3), a truncated form of ASBT or by the heteromeric organic solute (OST) α and β forms. Bile acids cause choleresis that is rich in bicarbonate ions secreted by the chloride/bicarbonate ion exchanger.

Unconjugated hydrophobic (mono- and dihydroxy) bile acids damage the bile-duct epithelium *in vitro* at concentrations as low as $10\,\mu M$ although damage became significant, particularly to mitochondria, at $50\,\mu M$. There appears to be no damage to the apical membrane or tight junctions, indicating that the damage reflects intra-cellular transport rather than a detergent effect at the apical membrane.[109] However, these authors also found that isolated livers perfused with bile acids failed to show damage to the bile-duct epithelium, even with depletion of taurine to increase the presence of unconjugated bile acids. This may reflect bile-acid binding during transport within the cholehepatic shunt, or perhaps the presence of phospholipids since cholangiocytes treated with $200\,\mu M$ conjugated bile acids went into apoptosis but PC prevented this in a concentration-dependent manner. The apoptosis was associated with enhanced ASBT levels but PC prevented this rise in ASBT and increased levels of MRP3 expression, leading to decreased uptake and enhanced basolateral secretion.[110]

2.13 Cholangiocyte Secretion

Cholangiocytes are responsible for the bile-acid-independent flow of bile that is regulated by a wide range of factors including the gastrointenstinal hormone secretin[111] and the peptide bombesin.[112] Secretin interacts with its receptor on cholangiocytes in large ducts, raising cAMP levels and activating the Cl^-/HCO_3^- exchanger that is essential for choleresis rich in bicarbonate.[113] Bile acids have been shown to increase secretin receptor gene expression *in vitro*[114] and *in vivo*[49] and this is probably the mechanism underlying the choleresis produced by bile acids.

Cholangiocytes express AQP1, which can be demonstrated at both the apical and basolateral membranes. Secretin promotion of bile flow also led to the microtubule-dependent insertion of AQP1 into the apical membrane but had no effect on levels in the basolateral membrane.[115] This group also reported the presence of AQP4 channels[116] that could explain the report by Mennone *et al.* who found that mice null for AQP1 showed no decrease in fluid secretion by bile-duct epithelia.[117] This is summarised in Figure 2.3.

2.14 Gallbladder

The gallbladder, lined with an epithelium of cells resembling cholangiocytes, serves as a storage system for bile during inter-digestive phases, effecting increased concentration of bile by active transport of Na^+, Cl^- and HCO_3^- ions leading to osmotically induced passive transport of water. This presumably involves aquaporins, as AQP1 and 8 have been reported in gallbladder, with AQP1 present at both apical and basolateral membranes and in vesicles around the apical membrane. AQP8 is found mainly within the apical membrane, with some present in intra-cellular vesicles.[118] This removal of water increases the concentration of bile acids, minimising the presence of monomeric bile acids that would enter the membrane bilayer and solubilise phospholipids into vesicles.[119]

In response to feeding, this fluid absorption is reversed and the epithelium secretes a bicarbonate-rich fluid similar to cholangiocyte secretion, also under the control of cAMP. The apical membrane expresses ASBT protein and uptake by this transporter regulates Cl^- ion extrusion and mucin secretion, similar to findings reported for cholangiocytes above. The Cl^- ion and mucin secretion is largely Na^+ dependent and may reflect a defence mechanism that protects against monomeric bile acids at the epithelium surface.[120] As bile acids potentiate cAMP-dependent Cl^- ion secretion, that in turn promote HCO_3^- and fluid secretion, this may indicate a bile-acid-induced mechanism that aids gallbladder emptying.[121] The hydrophobic bile acid taurochenodeoxycholate promotes greater mucin secretion than tauroursodeoxycholate[120] but similar effects on fluid secretion,[122] due to activation of PKC stimulating adenylyl cyclase and leading to increased cAMP levels.

Gallbladder contraction/motility has long been held to play a role in formation of gallstones in man. An investigation into 30 patients with cholesterol gallstones showed that gallbladder emptying could be used to divide the patients into strong and weak contractors, and weak contractors were associated with higher concentrations of total bile acids and phospholipids but this was unrelated to cholesterol concentration, bile-acid composition or specific species.[123] This higher concentration of bile acids and phospholipids could indicate reduced absorption by the gallbladder epithelium, or that the higher levels of bile acids in the lumen reduce motility. This study led to examination of muscle strips from gallbladder wall and showed that contraction in response to cholecystokinin or acetylcholine was reduced in the presence of either conjugated (μM) or unconjugated bile acids (10 μM) in a manner determined by the bile-acid hydrophobicity.[124] The *in-vivo* effect of bile acids on gallbladder emptying was confirmed when chenodeoxycholic acid infusion or oral administration reduced gallbladder emptying, while the hydrophilic bile acid ursodeoxycholic acid had no effect.[125]

2.15   Small Bowel

Gallbladder contraction ejects a bolus of bile acids into the small bowel where they participate in fat digestion.

2.16   Intestinal Fat Absorption

Pancreatic lipase is water soluble and acts at the surface of intestinal lipid droplets to hydrolise dietary triglycerides. Bile acids aid this process by their detergent properties that cause the formation of smaller lipid droplets and so increase surface area available for the lipase. The detergent properties are a result of the sterol nucleus that provides a hydrophobic surface with the α-hydroxyls providing a hydrophilic surface that is aided by ionisation of the conjugated carboxyl. Conjugated bile acids have pK values of 2 (taurine) and 4 (glycine) and, as the intestinal pH is normally in excess of 5, bile acids will be ionised. These detergent molecules form micelles above the critical micellar concentration and these micelles solubilise monoglycerides and fatty acids, the products of lipase action.

Bile acids have long been known to aid digestion of dietary fats, but are not essential. Some 50% of dietary fats are absorbed in rats where bile acids are diverted by biliary fistula.[126] Similar results were found in man.[127] This suggested that the micellar phase isolated by ultracentrifugation of duodenal contents was in fact composed of both bile-acid micelles and vesicles, a suggestion supported by a systematic study of the physical chemistry of fat digestion in human small bowel.[128]

2.17   Bile-Acid Absorption

Unconjugated bile acids have pK values of around 6, which means they will be un-ionised in the intestinal lumen and may be passively absorbed.[129] However, conjugated bile acids are ionised and require transporters to cross the enterocyte in much the same way as found in the hepatocyte. Although OATP2 is not expressed in small bowel a sodium-independent transporter is present in the

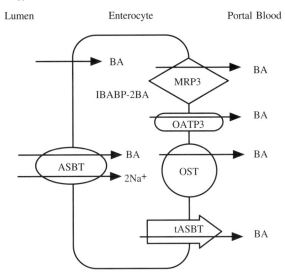

Figure 2.4 Bile acid absorption from the small bowel lumen. Bile acids are efficiently transported from the lumen of the terminal ileum by the apical sodium-dependent bile-acid transporter (ASBT). Unconjugated bile acids will be un-ionised at the pH of the lumen and may be passively absorbed. Within the enterocyte bile acids are bound by the intestinal bile-acid-binding protein (IBABP). Efflux from the enterocyte may involve the truncated ASBT (tASBT) and/or multi-drug-resistance-associated protein 3 (MRP3) but this remains to be defined. The heteromeric organic-solute transporter (OST) α and β gene products together but not separately transport bile acids out of the enterocyte. OATP3 mRNA has been identified but it is not yet clear whether the protein is functional.

brush border membranes of jejunum, identified as OATP3 when cloned initially from rat retina.[130] This was refined when a study found expression of OATP3 throughout the small bowel and identified the protein immunologically at the apical membrane.[131] These authors found weak but reproducible immunological staining, together with low levels of OATP3 mRNA suggesting low expression levels that would account for earlier studies failing to detect sodium-independent bile-acid transport. It has not yet been established that OATP3 is functional in small bowel and the Na^+-dependent transport of bile acids is the major pathway, although dysfunction of this pathway may up-regulate the Na^+-independent pathway. This is summarised in Figure 2.4.

2.17.1 Apical Sodium-Dependent Bile-Acid Transporter (ASBT)

This was originally cloned from a hamster cDNA library[132] but has since been cloned from a number of species including man,[133] rat[134] and mouse.[135] It is a 348 amino acid glycoprotein with either 7[136] or 9[137] transmembrane domains and with a cytosolic C terminal and extra-cellular N terminal.[138] The cytosolic

C terminal is required for apical targeting of the transporter as demonstrated by removal of 40 amino acids from the C terminal that prevented sorting to the apical membrane.[139]

In man, ASBT can transport unconjugated and conjugated bile acids but has a higher affinity for conjugates, and for dihydroxy over trihydroxy bile acids. This higher affinity for dihydroxy bile acids is associated with the orientation of hydroxyls as tauroursodeoxycholic acid (3α-,7β-dihydroxy) has a lower affinity than taurocholate (3α-, 7α-, 12α-trihydroxy).[140] Characterisation of rabbit ASBT has established a model for the binding of bile acids to this protein where the sterol 5-membered D ring with a methyl group at position 18 and α-hydroxyls at 7 or 12 are necessary for binding, although the presence of both decreases binding. The 3-hydroxyl orientation is not critical to binding. This accounts for the lower affinity and thus intestinal uptake of trihydroxy bile acids and for conjugated ursodeoxycholic acid where the 7-hydroxyl is β.[141]

Transport by ASBT is electrogenic with a 2:1 ratio of Na^+:bile acids and membrane potential may regulate transport function.[142] ASBT is essential for the enterohepatic circulation as shown by ASBT gene knockout mice that developed bile-acid malabsorption with no enterohepatic circulation.[143] This is summarised in Figure 2.4.

2.17.2 Regulation

ASBT has a complex regulatory system reflecting the importance of this transporter to bile-acid pool size and bile-acid synthesis rates. Hepatic nuclear factor 1α (HNF-1α) is necessary for expression of ASBT as knockout mice showed no expression and had defective bile-acid transport.[144] Conversely, FXR-null mice showed no difference in expression of ASBT,[64] showing that FXR plays no part in regulation of ASBT. In man, HNF-1α controls baseline promoter activity of the ASBT gene as the minimal construct with full promoter activity was found to have 3 HNF-1α binding sites.[145] These authors also showed that the promoter construct bound peroxisome proliferator activated receptor α (PPARα)/9 cis retinoic acid receptor heterodimer, demonstrating a link between bile-acid absorption and hepatic lipid metabolism mediated by PPARα.

Sterols also regulate ASBT by reducing both mRNA levels and promoter activity.[146] Feeding mice a diet with 2% cholesterol suppressed ileal uptake by 20% and increased faecal excretion of bile acids two-fold by down-regulation of ASBT expression. The authors confirmed their findings with isolated ileal loops (*ex vivo*) and transfected Caco2 cell lines (*in vitro*)[147] and suggest that regulation involves sterol regulatory element binding protein 2 (SREBP2), and HNF-1α acting co-operatively to regulate ASBT expression by an unknown mechanism. SREBP2 is functionally expressed in ileum as low levels of cholesterol promote cleavage of the N terminus to form the active or mature

SREBP2.[148] High levels of cholesterol would reduce levels of ASBT and reduce bile-acid absorption in the ileum.

The role of bile acids in regulation of ASBT is complicated by species differences, with no effect in rats[42] but showing negative feedback in mice[149] and humans.[150] This negative feedback is mediated by FXR and small heterodimer partner (SHP) causing repression of the RAR/RXR activation of ASBT.[150]

There is good evidence that inflammation of the intestine is associated with decreased activity of ASBT, as animal models show both reduced mRNA levels and a reduced affinity for bile acid.[151] Subsequently, interleukin 1β and tumour necrosis factor, pro-inflammatory cytokines, were shown to suppress the rat ASBT promoter in Caco2 cells by serine phosphorylation leading to nuclear translocation of c-fos. Similar results were found with indomethacin-treated rats leading the authors to suggest that ileal inflammation causes up-regulation and phosphorylation of c-fos, followed by its nuclear translocation causing reduced activity of ASBT and bile-acid malabsorption.[152] However, this picture may change in the presence of a full array of cytokines or using an experimental model of inflammation other than indomethacin, although supporting evidence was provided by ileal pinch biopsies taken from Crohn's patients that showed lower expression (69%) of ASBT.[153]

Hormones and vitamins also play a role in regulation of ASBT. Both glucocorticoid receptor ligands and co-expression of the glucocorticoid receptor gene increased activity of ASBT,[154] while there is also evidence that dihydroxy vitamin D binds directly to the vitamin D response element and increases expression of ASBT, leading to increased transport of bile acids into the enterocyte.[155]

2.17.3 Post-Translational Modification

In an analogous manner to BSEP, cAMP has been shown to increase activity of ileal ASBT,[156] This may be mediated by secretin that raises cAMP levels and has been shown to increase ASBT at the apical membrane of cholangiocytes by trafficking from a pool of sub-apical endosomes.[104] This would provide a rapid increase in ASBT activity, and there is evidence, again in cholangiocytes, to implicate the ubiquitin-proteosome system for rapid down-regulation of ASBT activity where the interleukin-1β effect was shown to be due to increased degradation following a serine/threonine phosphorylation.[157] Although there is no direct evidence of these mechanisms in the enterocyte the fact that cholangiocyte ASBT was cloned and found to be the same as the previously cloned ileal ASBT[103] makes it likely that regulation is similar.

2.18 Transport across the Enterocyte

As discussed for the hepatocyte it is unlikely that bile acids traverse the enterocyte as free monomers, and a binding protein has been identified.

Intestinal bile-acid-binding protein (IBABP) is a small 14–15-kDa protein found in the cytoplasm of cells in the ileum that binds bile acids as they enter the cell. By using photolabile bile-acid derivative and immunoprecipitation, expression was primarily found in the soluble protein fraction of ileal enterocytes, although cholangiocytes do show a low level of expression.[158] One molecule of IBABP binds two bile-acid molecules[159] as shown in Figure 2.4.

This protein functions in concert with ASBT to regulate bile-acid levels within the enterocyte and therefore expression is up-regulated by bile-acid flux through FXR activation, while ASBT expression is down-regulated. This was demonstrated in Caco2 cells and then in mice where sequestration of bile acids by cholestyramine caused a rapid decrease in mRNA for IBABP, while bile-acid feeding rapidly increased mRNA levels.[160] Transfection experiments in Caco2 cells showed that FXR and RXRα were both required for full activation by bile acids.[160] There are suggestions that cholesterol and PPAR are also involved in regulation of IBABP expression but the physiological relevance remains unclear and further studies are necessary to clarify this aspect of bile-acid transport across the enterocyte.

2.19   Export into Portal Blood

Efflux of bile acids from the enterocyte is mediated by an anion-exchange process as indicated in an early study with basolateral membrane vesicles from rat ileum that showed Na^+-independent taurocholate transport.[161] Some 14 years later the splice variant tASBT, reported to transport bile acids across the basolateral membrane in cholangiocytes, was found to be expressed at greater levels than the full-length version in ileal enterocytes.[107] However, MRP3, capable of bile-acid transport at the cholangiocyte basolateral membrane,[36] is also expressed in rat small bowel and colon[92] and human liver, small bowel and colon.[93] The presence of these transporters suggests they may have a role in the efflux of bile acids from the enterocyte, but this remains to be defined, particularly as a recent report showed no effect of ileal transport in MRP3-deficient mice.[37]

Recently, a new candidate protein has been identified, a heteromeric organic-solute transporter (OST) originally cloned from liver cDNA library of the skate.[162] Studies in man and mouse showed that two separate gene products, OSTα and OSTβ, expressed together but not independently, actively transport bile acids and some steroids.[163] Relatively high levels of OSTα/OSTβ are found at the basolateral membrane of enterocytes from the ileum and it is plausible that this heteromeric OST complex is responsible for efflux of bile acids into portal blood.[164] This is supported by the finding that exposure to chenodeoxycholic acid led to increased levels of OSTα/OSTβ mRNA, and gene expression was regulated by FXR,[165] showing similarity to FXR regulation of BSEP in hepatocytes. The transporters involved in bile-acid recovery from the ileal lumen and secretion into portal blood are shown in Figure 2.4.

2.20 Colon

Bile acids within the enterohepatic circulation that undergo absorption in the terminal ileum encounter a relatively low number of species and population of bacteria and return to the liver in portal blood relatively unchanged. However, the approximately 5% of the bile-acid pool that enters the colon provides substrate for the extensive microbial population that deconjugate and oxidise hydroxyl groups leading to formation of the secondary bile acids deoxycholic and lithocholic acids that are the major bile acids in faeces.

2.20.1 Deconjugation

Bile-acid hydrolases that cleave the amide bond linking C24 acyl group to glycine or taurine have been identified and cloned from a variety of species including *Clostridium perfringens,*[166] *Lactobacillus johnsonii,*[167] *Bifidobacterium bifidum*[168] and *Listeria monocytyogenes.*[169] Characterisation of the enzyme from *Bifidobacterium longum* showed a preferential hydrolysis of glycine residues and a marked sequence similarity with the penicillin V acylase found in *Bacillus sphaericus.*[170] This deconjugation provides glycine and taurine as substrates to provide cellular carbon, nitrogen and sulphur for the microbial flora. For the host, removal of the conjugated amino acid will raise the pK of the bile acid, limiting ionisation and allowing passive uptake of the nonionised species.

2.20.2 Oxidation of Hydroxyl Groups

Oxidation to form the keto group is reversible and may lead to epimerisation of the hydroxyl. The 3α/3β-hydroxy steroid dehydrogenases catalyse oxidation at the 3 position but there is a stereospecific preference for the α position as 3-oxo chenodeoxycholic acid is reduced to chenodeoxycholic acid (84%) by *Clostridium perfringens* rather than the 3β-hydroxyl.[171] Similar enzymes are specific for the 7 and 12 positions. Although the 7α/7β-hydoxy steroid dehydrogenases have been partially purified, less is known about their function as the irreversible removal of the 7α-hydroxyl complicates interpretation of studies in the intestine. Most of the species studied show greater affinity for dihydroxy bile acids that may be a protective action to minimise the antimicrobial activity of chenodeoxycholic acid, deoxycholic acid and lithocholic acid.[172] Although small amounts of β-hydoxy bile acids return to the liver after passive absorption only ursodeoxycholic acid (3α,7β) has significant concentration, as seen after chenodeoxycholic acid therapy.[173]

2.20.3 Dehydroxylation

The 7α-dehydroxylation is the most important bacterial transformation of bile acids, rapidly forming secondary from primary bile acids and is seemingly carried out by the *Clostridium* genus, around 0.0001% of colonic bacteria.[174] This dehydroxylation requires unconjugated bile acids, thus bile-acid hydrolase action is rapidly followed by 7α-dehydroxylation, converting cholic acid to deoxycholic acid and chenodeoxycholic acid to lithocholic acid.[175] These authors also reported that 7β-hydroxyls were poorly removed, suggesting that ursodeoxycholic acid would not be metabolised, although bacteria with 7α/7β-hydoxy steroid dehydrogenases could epimerise the 7β hydroxyl to 7α.

These deconjugated secondary bile acids are lipophilic and are believed to passively diffuse across the colon and enter the blood supply for return to the liver. Little is known of the mechanism, although in ASBT knockout mice there is an increase in OSTα/OSTβ mRNA within the proximal colon.[164] This could simply reflect reduced bile-acid uptake in the terminal ileum and a response to increased bile-acid levels entering the colon.

2.21   Measurement of Bile Acids

Assay of bile acids was an essential tool for the early investigation of the enterohepatic circulation, and proved a focus of attention with the belief that serum bile-acid concentrations would provide a sensitive diagnostic test for liver disease. There are three fundamental assay types, based on enzymatic oxidation of a hydroxyl with linked NAD reduction, chromatographic separations and quantitation, encompassing both gas-liquid and high-performance liquid chromatography, and radioimmunoassay assays.

2.21.1   Extraction

Most assay procedures require an extraction process to isolate and purify the bile acids as a minor component from biological fluids or tissue. The early methods used organic solvent partitioning techniques including the Folsch extraction[176] that involved homogenisation in chloroform:methanol $(2:1^v/_v)$ using some 20 ml/g, although optimal proportions should be determined empirically. This mixture is then centrifuged and the chloroform/methanol fraction removed and washed with 0.2 volumes of saline (0.9% NaCl). After further centrifugation the chloroform/methanol layer is evaporated to dryness at temperatures between room temperature and 45 °C. For larger volumes a rotary evaporator at reduced pressure can be used but for smaller analytical volumes the organic solvent is removed under a gentle nitrogen stream. Where the assay involved plasma, assay sensitivity determines the volume required; for normal samples this usually required 2 ml and proteins are precipitated with 3 ml of ethanol. This is left at 4 °C overnight to ensure precipitation of peptides. Next day the precipitate can be removed by centrifugation, washed with ethanol and the combined extract dried as described above.

More recent approaches have used resin extractions, initially the nonionic resin XAD2 and XAD7. These resins are effective where the pH was kept at 10 while mixing the plasma and resin. The supernatant is discarded and the bile acids then extracted three times from the resin with methanol. The methanol from the combined extracts is then dried under a nitrogen stream at 45 °C. These resins have been superseded by the bonded silica minicolumns packed with octadecylsilane (C18) although some analysts have used shorter sidechains (C8). The shorter sidechain seems to involve raised temperatures (64 °C) for the columns that are washed with aqueous methanol (65%) before elution with hexane/chloroform (95:5).[177] This is a more complex procedure than required for C18 columns.[178] These extraction columns are commercially available from suppliers such as Waters and Varian who provide details of procedures to prepare these columns for use. Elution is usually by methanol or ethanol, after which the solvent is removed.

2.21.2 Hydroxysteroid Dehydrogenases

The most common assay uses 3α-hydroxysteroid dehydrogenase to form the 3-keto bile acid that is trapped by, for example, hydrazine hydrate, causing the reaction to go to completion. The co-factor NAD is reduced stoichiometrically and can be measured by ultraviolet absorption or more commonly by fluorescence at an activation of 345 nm and emission of 450 nm. Use of this enzyme measures all bile acids with a 3α-hydroxyl but not cholesterol, which has a 3β-hydroxyl, and does not measure bile acids with a sulphate or glucuronide group conjugated to the 3α-hydroxyl.

Assays have also made use of 7α-hydroxysteroid dehydrogenase that can measure the primary bile acids, or for more specialised purposes such as differentiating between pathways of bile-acid synthesis to determine the proportion derived from the acid pathway.[178]

2.21.3 Chromatographic Assays

2.21.3.1 Gas-Liquid Chromatography (glc)

Gas-liquid chromatography requires several chemical processes after extraction before the analysis can be carried out. The first step involves the use of cholylglycine hydrolase to remove the conjugated amino acids from the carboxyl. This is now commercially available but originally was available only as an acetone powder of *Clostridium welchii* cells. This gave variable results and great care was required to establish the kinetics to ensure complete hydrolysis of the taurine conjugates. Unconjugated bile acids could be measured in samples by omitting the cholylglycine hydrolase. Some workers have removed the conjugated amino acids by alkaline hydrolysis using 2.5 M sodium hydroxide at 160 °C.[179] While this seems to be successful there are concerns that some bile-acid degradation occurs at the elevated alkaline conditions. After completion of the hydrolysis bile acids were extracted with a solvent such as ethyl acetate, dried and then methylated. The best reagent was diazomethane but borontrifluoride in methanol also provides quantitative methylation. Finally, the hydroxyls must be reacted to optimise chromatography. The derivative of choice was tri-fluoroacetic esters that were easy to prepare and gave a good linear range, although trimethylsilyl esters were also used. Originally packed columns with flame-ionisation detection were used and as the technology developed, capillary columns were adopted. Although the superior resolving power probably gave little real benefit to bile-acid analysis the enhanced column stability was welcome. The detection technology also developed, as mass spectroscopy detectors gave definitive identification of components. The level of detection with packed columns was at the limit for measurement of normal serum bile-acid concentrations, but the current use of electron-capture detection and mass-fragment detection has lowered this to 1 pg.

2.21.3.2   *High-Performance Liquid Chromatography (hplc)*

This method suffered from sensitivity problems initially as the bile-acid molecules lack a chromophore, but did offer the distinct advantage that conjugated bile acids could be determined without hydrolysis. The sensitivity issue was addressed by use of fluorescent derivatives such as dimethoxycoumarin esters[178] with a C18 reverse phase column and were able to resolve endogenous mixtures of bile acids. The combination of hplc and mass-spectroscopy detection has further improved the sensitivity along with providing specific identification, important as the resolution of bile acids by hplc is not as good as capillary column glc.[179]

2.21.4   Radioimmunoassay

The methods involving hplc and glc were too labour intensive to provide a routine method for diagnostic purposes, and the 3α-hydroxysteroid dehydrogenase enzyme method provided total bile-acid levels that hid potential diagnostic features believed to be available with individual bile-acid levels. This was resolved by using carbodiimide or mixed anhydride reactions to conjugate bile acids to albumin and then using that to raise antibodies against the bile-acid moiety. With careful definition of cross-reactivities of the anti-sera, specific and sensitive radioimmunoassay were established[180] and used widely to investigate diagnostic aspects of bile acids in liver disease.[181]

   The mixed anhydride method was used with diaminoethane to give a free amino group that was subsequently reacted with 4-hydroxybenzoic acid to provide an aromatic ring to allow iodination with [^{131}I] to increase sensitivity of the assay compared with labelled radioactive tracers.[182] This reaction could also be used to prepare fluorescent compounds for cell-biology studies.

Abbreviations

ABC   ATP-binding cassette
AQP   aquaporin

ASBT   apical sodium-dependent bile-acid transporter
ATP   adenosinetriphosphate
BAR   bile-acid receptor
BSEP   bile-salt export pump
cAMP   cyclic adenosinemonophosphate
cMOAT   canalicular multi-specific organic-anion transporter
dBcAMP dibutyrylcyclicadenosinemonophosphate
FXR   farnesyl X receptor
Glc   gas-liquid chromatography
HNF   hepatic nuclear factor
Hplc   high-performance liquid chromatography
IBABP   intestinal bile-acid-binding protein
MDR   multi-drug resistance
MEH   microsomal epoxide hydrolase
MRP   multi-drug-resistance-associated protein
NTCP   sodium taurocholate co-transporting polypeptide
OATP   organic-anion-transporting polypeptide
OST   organic-solute transporter
PC   phosphatidylcholine
PI   phosphatidylinositol
PKA   protein kinase A
PKC   protein kinase C
PPAR   peroxisome proliferator activated receptor
RAR   retinoic acid receptor
RXR   retinoid X receptor
SREBP   sterol regulatory element binding protein
tASBT   truncated apical sodium-dependent bile-acid transporter

Proteins from animal species are usually designated with the first letter upper case and subsequent letters in lower case, while those from man are all in upper case. For simplicity all upper case is used generically unless an animal source is described.

References

1. A. Reuben, *Hepatology*, 2005, **42**, 500.
2. W. Kramer, H. P. Buscher, W. Gerok and G. Kurz, *Eur. J. Biochem.*, 1979, **102**, 1.
3. A. Roda, G. Cappelleri, R. Aldini, E. Roda and L. Barbara, *J. Lipid Res.*, 1982, **23**, 490.
4. L. R. Schwarz, R. Burr, M. Schwenk, E. Pfaff and H. Greim, *Eur. J. Biochem.*, 1975, **55**, 617.
5. J. Reichen and G. Paumgartner, *Am. J. Physiol.*, 1976, **231**, 734.
6. P. von Dippe, P. Drain and D. Levy, *J. Biol. Chem.*, 1983, **258**, 8890.
7. P. J. Meier, *Am. J. Physiol.*, 1995, **269**, G801.

8. A. Schroeder, U. Eckhardt, B. Stieger, R. Tynes, C. D. Schteingart, A. F. Hofmann, P. J. Meier and B. Hagenbuch, *Am. J. Physiol.*, 1998, **274**, G370.
9. S. A. Weinman, *Yale J. Biol. Med.*, 1997, **70**, 331.
10. D. Liang, B. Hagenbuch, B. Stieger and P. J. Meier, *Hepatology*, 1993, **18**, 1162.
11. F. J. Suchy, J. C. Bucuvalas, A. L. Goodrich, M. S. Moyer and B. L. Blitzer, *Am. J. Physiol.*, 1986, **251**, G665.
12. B. Hagenbuch, B. F. Scharschmidt and P. J. Meier, *Biochem. J.*, 1996, **316** (Pt 3), 901.
13. A. W. Wolkoff and D. E. Cohen, *Am. J. Physiol. Gastrointest Liver Physiol*, 2003, **284**, G175.
14. B. Hagenbuch, B. Stieger, M. Foguet, H. Lubbert and P. J. Meier, *Proc. Natl. Acad. Sci. USA*, 1991, **88**, 10629.
15. V. Cattori, U. Eckhardt and B. Hagenbuch, *Biochim. Biophys. Acta.*, 1999, **1445**, 154.
16. W. Kramer, S. Stengelin, K. H. Baringhaus, A. Enhsen, H. Heuer, W. Becker, D. Corsiero, F. Girbig, R. Noll and C. Weyland, *J. Lipid Res.*, 1999, **40**, 1604.
17. B. Hagenbuch and P. J. Meier, *J. Clin. Invest.*, 1994, **93**, 1326.
18. P. J. Meier, U. Eckhardt, A. Schroeder, B. Hagenbuch and B. Stieger, *Hepatology*, 1997, **26**, 1667.
19. M. Ananthanarayanan, O. C. Ng, J. L. Boyer and F. J. Suchy, *Am. J. Physiol.*, 1994, **267**, G637.
20. B. Stieger, B. Hagenbuch, L. Landmann, M. Hochli, A. Schroeder and P. J. Meier, *Gastroenterology*, 1994, **107**, 1781.
21. P. von Dippe, M. Amoui, R. H. Stellwagen and D. Levy, *J. Biol. Chem.*, 1996, **271**, 18176.
22. C. Alves, P. von Dippe, M. Amoui and D. Levy, *J. Biol. Chem.*, 1993, **268**, 20148.
23. E. Jacquemin, B. Hagenbuch, B. Stieger, A. W. Wolkoff and P. J. Meier, *Proc. Natl. Acad. Sci. USA*, 1994, **91**, 133.
24. B. Hagenbuch and P. J. Meier, *Biochim. Biophys. Acta.*, 2003, **1609**, 1.
25. L. M. Satlin, V. Amin and A. W. Wolkoff, *J. Biol. Chem.*, 1997, **272**, 26340.
26. L. Li, T. K. Lee, P. J. Meier and N. Ballatori, *J. Biol. Chem.*, 1998, **273**, 16184.
27. W. Kramer, S. Stengelin, K.-H. Baringhaus, A. Enhsen, H. Heuer, W. Becker, D. Corsiero, F. Girbig, R. Noll and C. Weyland, *J. Lipid Res.*, 1999, **40**, 1604.
28. S. Hata, P. Wang, N. Eftychiou, M. Ananthanarayanan, A. Batta, G. Salen, K. S. Pang and A. W. Wolkoff, *Am. J. Physiol. Gastrointest Liver Physiol.*, 2003, **285**, G829.
29. E. C. Friesema, R. Docter, E. P. Moerings, B. Stieger, B. Hagenbuch, P. J. Meier, E. P. Krenning, G. Hennemann and T. J. Visser, *Biochem. Biophys. Res. Commun.*, 1999, **254**, 497.

30. Y. Cui, J. Konig, I. Leier, U. Buchholz and D. Keppler, *J. Biol. Chem.*, 2001, **276**, 9626.
31. A. D. Min, K. L. Johansen, C. G. Campbell and A. W. Wolkoff, *J. Clin. Invest.*, 1991, **87**, 1496.
32. A. W. Wolkoff, A. C. Samuelson, K. L. Johansen, R. Nakata, D. M. Withers and A. Sosiak, *J. Clin. Invest.*, 1987, **79**, 1259.
33. W. Honscha, H. D. Platte, F. Oesch and T. Friedberg, *Biochem. J.*, 1995, **311** (Pt 3), 975.
34. M. Miyata, G. Kudo, Y. H. Lee, T. J. Yang, H. V. Gelboin, P. Fernandez-Salguero, S. Kimura and F. J. Gonzalez, *J. Biol. Chem.*, 1999, **274**, 23963.
35. H. Zeng, G. Liu, P. A. Rea and G. D. Kruh, *Cancer Res.*, 2000, **60**, 4779.
36. C. J. Soroka, J. M. Lee, F. Azzaroli and J. L. Boyer, *Hepatology*, 2001, **33**, 783.
37. N. Zelcer, K. van de Wetering, R. de Waart, G. L. Scheffer, H. U. Marschall, P. R. Wielinga, A. Kuil, C. Kunne, A. Smith, M. van der Valk, J. Wijnholds, R. O. Elferink and P. Borst, *J. Hepatol.*, 2006, **44**, 768.
38. S. J. Karpen, A. Q. Sun, B. Kudish, B. Hagenbuch, P. J. Meier, M. Ananthanarayanan and F. J. Suchy, *J. Biol. Chem.*, 1996, **271**, 15211.
39. D. Jung, B. Hagenbuch, M. Fried, P. J. Meier and G. A. Kullak-Ublick, *Am. J. Physiol. Gastrointest Liver Physiol.*, 2004, **286**, G752.
40. M. S. Anwer, H. Gillin, S. Mukhopadhyay, N. Balasubramaniyan, F. J. Suchy and M. Ananthanarayanan, *J. Biol. Chem.*, 2005, **280**, 33687.
41. G. L. Guo and C. D. Klaassen, *J. Pharmacol. Exp. Ther.*, 2001, **299**, 551.
42. M. Arrese, M. Trauner, R. J. Sacchiero, M. W. Crossman and B. L. Shneider, *Hepatology*, 1998, **28**, 1081.
43. F. J. Suchy, W. F. Balistreri, J. Hung, P. Miller and S. A. Garfield, *Am. J. Physiol. Gastrointest Liver Physiol.*, 1983, **245**, G681.
44. F. A. Simion, B. Fleischer and S. Fleischer, *Biochemistry*, 1984, **23**, 6459.
45. A. Z. El-Seaidy, C. O. Mills, E. Elias and J. M. Crawford, *Am. J. Physiol. Gastrointest Liver Physiol.*, 1997, **272**, G298.
46. J. M. Crawford, Y. J. Lin, B. A. Teicher, J. P. Narciso and J. L. Gollan, *Biochim. Biophys. Acta.*, 1991, **1085**, 223.
47. A. Dietrich, W. Dieminger, K. Fuchte, G. H. Stoll, E. Schlitz, W. Gerok and G. Kurz, *J. Lipid Res.*, 1995, **36**, 1745.
48. A. E. Thumser and D. C. Wilton, *Biochem. J.*, 1996, **320**, 729.
49. H. Takikawa, A. Stoltz and N. Kaplowitz, *J. Clin. Invest.*, 1987, **80**, 852.
50. H. Takikawa, M. Ookhtens, A. Stolz and N. Kaplowitz, *J. Clin. Invest.*, 1987, **80**, 861.
51. A. Stolz, L. Hammond, H. Lou, H. Takikawa, M. Ronk and J. E. Shively, *J. Biol. Chem.*, 1993, **268**, 10448.
52. H. Nanjo, H. Adachi, S. Morihana, T. Mizoguchi, T. Nishihara and T. Terada, *Biochim. Biophys. Acta.*, 1995, **1244**, 53.
53. K. Solaas, A. Ulvestad, O. Soreide and B. F. Kase, *J. Lipid Res.*, 2000, **41**, 1154.

54. C. J. Sippel, M. Ananthanarayanan and F. J. Suchy, *Am. J. Physiol. Gastrointest Liver Physiol.*, 1990, **258**, G728.
55. C. J. Sippel, F. J. Suchy, M. Ananthanarayanan and D. H. Perlmutter, *J. Biol. Chem.*, 1993, **268**, 2083.
56. C. Kast, B. Stieger, K. H. Winterhalter and P. J. Meier, *J. Biol. Chem.*, 1994, **269**, 5179.
57. S. Childs, R. L. Yeh, E. Georges and V. Ling, *Cancer Res.*, 1995, **55**, 2029.
58. R. S. Brown Jr., N. Lomri, J. De Voss, C. M. Rahmaoui, M. H. Xie, T. Hua, S. D. Lidofsky and B. F. Scharschmidt, *Proc. Natl. Acad. Sci. USA*, 1995, **92**, 5421.
59. T. Gerloff, B. Stieger, B. Hagenbuch, J. Madon, L. Landmann, J. Roth, A. F. Hofmann and P. J. Meier, *J. Biol. Chem.*, 1998, **273**, 10046.
60. B. Stieger, K. Fattinger, J. Madon, G. A. Kullak-Ublick and P. J. Meier, *Gastroenterology*, 2000, **118**, 422.
61. P. L. Jansen, S. S. Strautnieks, E. Jacquemin, M. Hadchouel, E. M. Sokal, G. J. Hooiveld, J. H. Koning, A. De Jager-Krikken, F. Kuipers, F. Stellaard, C. M. Bijleveld, A. Gouw, H. Van Goor, R. J. Thompson and M. Muller, *Gastroenterology*, 1999, **117**, 1370.
62. R. Wang, M. Salem, I. M. Yousef, B. Tuchweber, P. Lam, S. J. Childs, C. D. Helgason, C. Ackerley, M. J. Phillips and V. Ling, *Proc. Natl. Acad. Sci. USA*, 2001, **98**, 2011.
63. P. Lam, R. Wang and V. Ling, *Biochemistry*, 2005, **44**, 12598.
64. C. J. Sinal, M. Tohkin, M. Miyata, J. M. Ward, G. Lambert and F. J. Gonzalez, *Cell*, 2000, **102**, 731.
65. M. Ananthanarayanan, N. Balasubramanian, M. Makishima, D. J. Mangelsdorf and F. J. Suchy, *J. Biol. Chem.*, 2001, **276**, 28857.
66. S. Misra, P. Ujhazy, Z. Gatmaitan, L. Varticovski and I. M. Arias, *J. Biol. Chem.*, 1998, **273**, 26638.
67. Z. C. Gatmaitan, A. T. Nies and I. M. Arias, *Am. J. Physiol.*, 1997, **272**, G1041.
68. S. Misra, P. Ujhazy, L. Varticovski and I. M. Arias, *Proc. Natl. Acad. Sci. USA*, 1999, **96**, 5814.
69. K. M. Botham and K. E. Suckling, *Biochim. Biophys. Acta.*, 1986, **883**, 26.
70. S. Misra, L. Varticovski and I. M. Arias, *Am. J. Physiol. Gastrointest Liver Physiol.*, 2003, **285**, G316.
71. T. Kagawa, L. Varticovski, Y. Sai and I. M. Arias, *Am. J. Physiol. Cell Physiol.*, 2002, **283**, C1655.
72. P. Wang, R. B. Kim, J. R. Chowdhury and A. W. Wolkoff, *J. Biol. Chem.*, 2003, **278**, 20695.
73. H. Roelofsen, C. J. Soroka, D. Keppler and J. L. Boyer, *J. Cell Sci.*, 1998, **111** (Pt 8), 1137.
74. A. Benedetti, M. Strazzabosco, O. C. Ng and J. L. Boyer, *Proc. Natl. Acad. Sci. USA*, 1994, **91**, 792.
75. T. Hayakawa, R. Bruck, O. C. Ng and J. L. Boyer, *Am. J. Physiol.*, 1990, **259**, G727.

76. M. M. Zegers and D. Hoekstra, *J. Cell Biol.*, 1997, **138**, 307.
77. F. A. Crocenzi, A. D. Mottino, J. Cao, L. M. Veggi, E. J. Pozzi, M. Vore, R. Coleman and M. G. Roma, *Am. J. Physiol. Gastrointest Liver Physiol.*, 2003, **285**, G449.
78. R. Bossard, B. Stieger, B. O'Neill, G. Fricker and P. J. Meier, *J. Clin. Invest.*, 1993, **91**, 2714.
79. J. Noe, B. Hagenbuch, P. J. Meier and M. V. St-Pierre, *Hepatology*, 2001, **33**, 1223.
80. B. A. Jones, Y. P. Rao, R. T. Stravitz and G. J. Gores, *Am. J. Physiol.*, 1997, **272**, G1109.
81. U. Beuers, D. C. Throckmorton, M. S. Anderson, C. M. Isales, W. Thasler, G. A. Kullak-Ublick, G. Sauter, H. G. Koebe, G. Paumgartner and J. L. Boyer, *Gastroenterology*, 1996, **110**, 1553.
82. U. Beuers, M. Bilzer, A. Chittattu, G. A. Kullak-Ublick, D. Keppler, G. Paumgartner and F. Dombrowski, *Hepatology*, 2001, **33**, 1206.
83. P. M. Gerk and M. Vore, *J. Pharmacol. Exp. Ther.*, 2002, **302**, 407.
84. A. Lindahl, A. Sjoberg, U. Bredberg, H. Toreson, A. L. Ungell and H. Lennernas, *Mol. Pharm.*, 2004, **1**, 347.
85. K. Ito, H. Suzuki and Y. Sugiyama, *Am. J. Physiol. Gastrointest Liver Physiol.*, 2001, **281**, G1034.
86. S. Teng and M. Piquette-Miller, *Br. J. Pharmacol.*, 2007, **151**, 367.
87. M. H. Nathanson and J. L. Boyer, *Hepatology*, 1991, **14**, 551.
88. R. Coleman, S. Iqbal, P. P. Godfrey and D. Billington, *Biochem. J.*, 1979, **178**, 201.
89. R. Coleman, K. Rahman, K. S. Kan and R. A. Parslow, *Biochem. J.*, 1989, **258**, 17.
90. D. W. Hay, M. J. Cahalane, N. Timofeyeva and M. C. Carey, *J. Lipid Res.*, 1993, **34**, 759.
91. K. W. Wirtz, *Annu. Rev. Biochem.*, 1991, **60**, 73.
92. T. Hirohashi, H. Suzuki, K. Ito, K. Ogawa, K. Kume, T. Shimizu and Y. Sugiyama, *Mol. Pharmacol.*, 1998, **53**, 1068.
93. Y. Kiuchi, H. Suzuki, T. Hirohashi, C. A. Tyson and Y. Sugiyama, *FEBS Lett.*, 1998, **433**, 149.
94. J. J. Smit, A. H. Schinkel, R. P. Oude Elferink, A. K. Groen, E. Wagenaar, L. van Deemter, C. A. Mol, R. Ottenhoff, N. M. van der Lugt and M. A. van Roon, *et al. Cell*, 1993, **75**, 451.
95. J. M. de Vree, E. Jacquemin, E. Sturm, D. Cresteil, P. J. Bosma, J. Aten, J. F. Deleuze, M. Desrochers, M. Burdelski, O. Bernard, R. P. Oude Elferink and M. Hadchouel, *Proc. Natl. Acad. Sci. USA*, 1998, **95**, 282.
96. R. P. Oude Elferink, R. Ottenhoff, M. van Wijland, J. J. Smit, A. H. Schinkel and A. K. Groen, *J. Clin. Invest.*, 1995, **95**, 31.
97. A. van Helvoort, A. de Brouwer, R. Ottenhoff, J. F. H. M. Brouwers, J. Wijnholds, J. H. Beijnen, A. Rijneveld, T. van der Poll, M. A. van der Valk, D. Majoor, W. Voorhout, K. W. A. Wirtz, R. P. J. O. Elferink and P. Borst, *Proceedings of the National Academy of Sciences*, 1999, **96**, 11501.

98. E. R. Eckhardt, A. Moschetta, W. Renooij, S. S. Goerdayal, G. P. van Berge-Henegouwen and K. J. van Erpecum, *J. Lipid Res.*, 1999, **40**, 2022.

99. M. Yano, R. A. Marinelli, S. K. Roberts, V. Balan, L. Pham, J. E. Tarara, P. C. de Groen and N. F. LaRusso, *J. Biol. Chem.*, 1996, **271**, 6702.

100. R. A. Marinelli, P. S. Tietz, A. J. Caride, B. Q. Huang and N. F. LaRusso, *J. Biol. Chem.*, 2003, **278**, 43157.

101. R. C. Huebert, P. L. Splinter, F. Garcia, R. A. Marinelli and N. F. LaRusso, *J. Biol. Chem.*, 2002, **277**, 22710.

102. F. I. Carreras, G. L. Lehmann, D. Ferri, M. F. Tioni, G. Calamita and R. A. Marinelli, *Am. J. Physiol. Gastrointest Liver Physiol.*, 2007, **292**, G905.

103. K. N. Lazaridis, L. Pham, P. Tietz, R. A. Marinelli, P. C. deGroen, S. Levine, P. A. Dawson and N. F. LaRusso, *J. Clin. Invest.*, 1997, **100**, 2714.

104. G. Alpini, S. Glaser, L. Baiocchi, H. Francis, X. Xia and G. Lesage, *Hepatology*, 2005, **41**, 1037.

105. G. Alpini, S. Glaser, D. Alvaro, Y. Ueno, M. Marzioni, H. Francis, L. Baiocchi, T. Stati, B. Barbaro, J. L. Phinizy, J. Mauldin and G. Lesage, *Gastroenterology*, 2002, **123**, 1226.

106. N. W. Werneburg, J. H. Yoon, H. Higuchi and G. J. Gores, *Am. J. Physiol. Gastrointest Liver Physiol.*, 2003, **285**, G31.

107. K. N. Lazaridis, P. Tietz, T. Wu, S. Kip, P. A. Dawson and N. F. LaRusso, *Proc. Natl. Acad. Sci. USA*, 2000, **97**, 11092.

108. X. Xia, H. Francis, S. Glaser, G. Alpini and G. LeSage, *World J. Gastroenterol*, 2006, **12**, 3553.

109. A. Benedetti, D. Alvaro, C. Bassotti, A. Gigliozzi, G. Ferretti, T. La Rosa, A. Di Sario, L. Baiocchi and A. M. Jezequel, *Hepatology*, 1997, **26**, 9.

110. K. Tsuboi, S. Tazuma, T. Nishioka and K. Chayama, *J. Gastroenterol*, 2004, **39**, 955.

111. G. Alpini, R. Lenzi, L. Sarkozi and N. Tavoloni, *J. Clin. Invest.*, 1988, **81**, 569.

112. W. K. Cho, A. Mennone, S. A. Rydberg and J. L. Boyer, *Gastroenterology*, 1997, **113**, 311.

113. D. Alvaro, G. Alpini, A. M. Jezequel, C. Bassotti, C. Francia, F. Fraioli, R. Romeo, L. Marucci, G. Le Sage, S. S. Glaser and A. Benedetti, *J. Clin. Invest.*, 1997, **100**, 1349.

114. G. Alpini, S. Glaser, W. Robertson, J. L. Phinizy, R. E. Rodgers, A. Caligiuri and G. LeSage, *Am. J. Physiol. Gastrointest Liver Physiol.*, 1997, **273**, G518.

115. R. A. Marinelli, P. S. Tietz, L. D. Pham, L. Rueckert, P. Agre and N. F. LaRusso, *Am. J. Physiol.*, 1999, **276**, G280.

116. R. A. Marinelli, L. D. Pham, P. S. Tietz and N. F. LaRusso, *Hepatology*, 2000, **31**, 1313.

117. A. Mennone, A. S. Verkman and J. L. Boyer, *Am. J. Physiol. Gastrointest Liver Physiol.*, 2002, **283**, G739.

118. A. I. Masyuk and N. F. LaRusso, *Hepatology*, 2006, **43**, S75.

119. D. A. Fahey, M. C. Carey and J. M. Donovan, *Biochemistry*, 1995, **34**, 10886.

120. N. Chignard, M. Mergey, D. Veissiere, R. Parc, J. Capeau, R. Poupon, A. Paul and C. Housset, *Hepatology*, 2001, **33**, 496.

121. T. Chinet, L. Fouassier, N. Dray-Charier, M. Imam-Ghali, H. Morel, M. Mergey, B. Dousset, R. Parc, A. Paul and C. Housset, *Hepatology*, 1999, **29**, 5.

122. N. Chignard, M. Mergey, D. Veissiere, R. Poupon, J. Capeau, R. Parc, A. Paul and C. Housset, *Am. J. Physiol. Gastrointest Liver Physiol.*, 2003, **284**, G205.

123. M. F. Stolk, K. J. van Erpecum, W. Renooij, P. Portincasa, B. J. van de Heijning and G. P. vanBerge-Henegouwen, *Gastroenterology*, 1995, **108**, 1882.

124. M. F. Stolk, B. J. Van de Heijning, K. J. Van Erpecum, A. Verheem, L. M. Akkermans and G. P. Van Berge-Henegouwen, *Ital. J. Gastroenterol*, 1996, **28**, 105.

125. N. Carulli, M. Bertolotti, F. Carubbi, M. Concari, P. Martella, L. Carulli and P. Loria, *Aliment Pharmacol. Ther.*, 2000, **14** (Suppl 2), 14.

126. N. Gallagher, J. Webb and A. M. Dawson, *Clin. Sci.*, 1965, **29**, 73.

127. H. P. Porter, D. R. Saunders, G. Tytgat, O. Brunser and C. E. Rubin, *Gastroenterology*, 1971, **60**, 1008.

128. O. Hernell, J. E. Staggers and M. C. Carey, *Biochemistry*, 1990, **29**, 2041.

129. R. J. Bahar and A. Stolz, *Gastroenterol Clin. North Am.*, 1999, **28**, 27.

130. T. Abe, M. Kakyo, H. Sakagami, T. Tokui, T. Nishio, M. Tanemoto, H. Nomura, S. C. Hebert, S. Matsuno, H. Kondo and H. Yawo, *J. Biol. Chem.*, 1998, **273**, 22395.

131. H. C. Walters, A. L. Craddock, H. Fusegawa, M. C. Willingham and P. A. Dawson, *Am. J. Physiol. Gastrointest Liver Physiol.*, 2000, **279**, G1188.

132. M. H. Wong, P. Oelkers, A. L. Craddock and P. A. Dawson, *J. Biol. Chem.*, 1994, **269**, 1340.

133. M. H. Wong, P. Oelkers and P. A. Dawson, *J. Biol. Chem.*, 1995, **270**, 27228.

134. B. L. Shneider, P. A. Dawson, D. M. Christie, W. Hardikar, M. H. Wong and F. J. Suchy, *J. Clin. Invest.*, 1995, **95**, 745.

135. T. Saeki, K. Matoba, H. Furukawa, K. Kirifuji, R. Kanamoto and K. Iwami, *J. Biochem. (Tokyo)*, 1999, **125**, 846.

136. A. Banerjee and P. W. Swaan, *Biochemistry*, 2006, **45**, 943.

137. S. Hallen, M. Branden, P. A. Dawson and G. Sachs, *Biochemistry*, 1999, **38**, 11379.

138. B. Hagenbuch and P. Dawson, *Pflugers Arch.*, 2004, **447**, 566.

139. A. Q. Sun, M. Ananthanarayanan, C. J. Soroka, S. Thevananther, B. L. Shneider and F. J. Suchy, *Am. J. Physiol.*, 1998, **275**, G1045.

140. A. L. Craddock, M. W. Love, R. W. Daniel, L. C. Kirby, H. C. Walters, M. H. Wong and P. A. Dawson, *Am. J. Physiol.*, 1998, **274**, G157.

141. K. H. Baringhaus, H. Matter, S. Stengelin and W. Kramer, *J. Lipid Res.*, 1999, **40**, 2158.

142. S. A. Weinman, M. W. Carruth and P. A. Dawson, *J. Biol. Chem.*, 1998, **273**, 34691.
143. P. A. Dawson, J. Haywood, A. L. Craddock, M. Wilson, M. Tietjen, K. Kluckman, N. Maeda and J. S. Parks, *J. Biol. Chem.*, 2003, **278**, 33920.
144. D. Q. Shih, M. Bussen, E. Sehayek, M. Ananthanarayanan, B. L. Shneider, F. J. Suchy, S. Shefer, J. S. Bollileni, F. J. Gonzalez, J. L. Breslow and M. Stoffel, *Nat. Genet.*, 2001, **27**, 375.
145. D. Jung, M. Fried and G. A. Kullak-Ublick, *J. Biol. Chem.*, 2002, **277**, 30559.
146. W. A. Alrefai, Z. Sarwar, S. Tyagi, S. Saksena, P. K. Dudeja and R. K. Gill, *Am. J. Physiol. Gastrointest Liver Physiol.*, 2005, **288**, G978.
147. C. Thomas, J. F. Landrier, D. Gaillard, J. Grober, M. C. Monnot, A. Athias and P. Besnard, *Gut*, 2006, **55**, 1321.
148. J. D. Horton, J. L. Goldstein and M. S. Brown, *J. Clin. Invest.*, 2002, **109**, 1125.
149. A. Figge, F. Lammert, B. Paigen, A. Henkel, S. Matern, R. Korstanje, B. L. Shneider, F. Chen, E. Stoltenberg, K. Spatz, F. Hoda, D. E. Cohen and R. M. Green, *J. Biol. Chem.*, 2004, **279**, 2790.
150. E. Neimark, F. Chen, X. Li and B. L. Shneider, *Hepatology*, 2004, **40**, 149.
151. U. Sundaram, S. Wisel, S. Stengelin, W. Kramer and V. Rajendran, *Am. J. Physiol.*, 1998, **275**, G1259.
152. F. Chen, L. Ma, R. B. Sartor, F. Li, H. Xiong, A. Q. Sun and B. Shneider, *Gastroenterology*, 2002, **123**, 2005.
153. D. Jung, A. C. Fantin, U. Scheurer, M. Fried and G. A. Kullak-Ublick, *Gut*, 2004, **53**, 78.
154. M. J. Nowicki, B. L. Shneider, J. M. Paul and J. E. Heubi, *Am. J. Physiol.*, 1997, **273**, G197.
155. J. L. Boyer, M. Trauner, A. Mennone, C. J. Soroka, S. Y. Cai, T. Moustafa, G. Zollner, J. Y. Lee and N. Ballatori, *Am. J. Physiol. Gastrointest Liver Physiol.*, 2006, **290**, G1124.
156. A. Reymann, W. Braun, C. Drobik and C. Woermann, *Biochim. Biophys. Acta.*, 1989, **1011**, 158.
157. X. Xia, M. Roundtree, A. Merikhi, X. Lu, S. Shentu and G. Lesage, *J. Biol. Chem.*, 2004, **279**, 44931.
158. M. C. Lin, W. Kramer and F. A. Wilson, *J. Biol. Chem.*, 1990, **265**, 14986.
159. O. Toke, J. D. Monsey, G. T. DeKoster, G. P. Tochtrop, C. Tang and D. P. Cistola, *Biochemistry*, 2006, **45**, 727.
160. J. Grober, I. Zaghini, H. Fujii, S. A. Jones, S. A. Kliewer, T. M. Willson, T. Ono and P. Besnard, *J. Biol. Chem.*, 1999, **274**, 29749.
161. S. L. Weinberg, G. Burckhardt and F. A. Wilson, *J. Clin. Invest.*, 1986, **78**, 44.
162. W. Wang, D. J. Seward, L. Li, J. L. Boyer and N. Ballatori, *Proc. Natl. Acad. Sci. USA*, 2001, **98**, 9431.
163. D. J. Seward, A. S. Koh, J. L. Boyer and N. Ballatori, *J. Biol. Chem.*, 2003, **278**, 27473.

164. P. A. Dawson, M. Hubbert, J. Haywood, A. L. Craddock, N. Zerangue, W. V. Christian and N. Ballatori, *J. Biol. Chem.*, 2005, **280**, 6960.
165. J. F. Landrier, J. J. Eloranta, S. R. Vavricka and G. A. Kullak-Ublick, *Am. J. Physiol. Gastrointest Liver Physiol.*, 2006, **290**, G476.
166. J. P. Coleman and L. L. Hudson, *Appl. Environ. Microbiol.*, 1995, **61**, 2514.
167. C. A. Elkins and D. C. Savage, *J. Bacteriol.*, 1998, **180**, 4344.
168. G. B. Kim, C. M. Miyamoto, E. A. Meighen and B. H. Lee, *Appl. Environ. Microbiol.*, 2004, **70**, 5603.
169. P. Glaser, L. Frangeul, C. Buchrieser, C. Rusniok, A. Amend, F. Baquero, P. Berche, H. Bloecker, P. Brandt, T. Chakraborty, A. Charbit, F. Chetouani, E. Couve, A. de Daruvar, P. Dehoux, E. Domann, G. Dominguez-Bernal, E. Duchaud, L. Durant, O. Dussurget, K. D. Entian, H. Fsihi, F. Garcia-del Portillo, P. Garrido, L. Gautier, W. Goebel, N. Gomez-Lopez, T. Hain, J. Hauf, D. Jackson, L. M. Jones, U. Kaerst, J. Kreft, M. Kuhn, F. Kunst, G. Kurapkat, E. Madueno, A. Maitournam, J. M. Vicente, E. Ng, H. Nedjari, G. Nordsiek, S. Novella, B. de Pablos, J. C. Perez-Diaz, R. Purcell, B. Remmel, M. Rose, T. Schlueter, N. Simoes, A. Tierrez, J. A. Vazquez-Boland, H. Voss, J. Wehland and P. Cossart, *Science*, 2001, **294**, 849.
170. H. Tanaka, H. Hashiba, J. Kok and I. Mierau, *Appl. Environ. Microbiol.*, 2000, **66**, 2502.
171. I. A. Macdonald, D. M. Hutchison, T. P. Forrest, V. D. Bokkenheuser, J. Winter and L. V. Holdeman, *J. Steroid. Biochem.*, 1983, **18**, 97.
172. M. Begley, C. G. Gahan and C. Hill, *FEMS Microbiol. Rev.*, 2005, **29**, 625.
173. G. Salen, G. S. Tint, B. Eliav, N. Deering and E. H. Mosbach, *J. Clin. Invest.*, 1974, **53**, 612.
174. E. J. Stellwag and P. B. Hylemon, *Am. J. Clin. Nutr.*, 1978, **31**, S243.
175. A. K. Batta, G. Salen, R. Arora, S. Shefer, M. Batta and A. Person, *J. Biol. Chem.*, 1990, **265**, 10925.
176. J. Folch, M. Lees and G. H. Sloane Stanley, *J. Biol. Chem.*, 1957, **226**, 497.
177. C. Galman, I. Arvidsson, B. Angelin and M. Rudling, *J. Lipid Res.*, 2003, **44**, 859.
178. K. Budai and N. B. Javitt, *J. Lipid Res.*, 1997, **38**, 1906.
179. S. Perwaiz, B. Tuchweber, D. Mignault, T. Gilat and I. M. Yousef, *J. Lipid Res.*, 2001, **42**, 114.
180. Y. A. Baqir, J. Murison, P. E. Ross and I. A. Bouchier, *J. Clin. Pathol.*, 1979, **32**, 560.
181. P. E. Ross, *Methods Enzymol.*, 1982, **84**, 321.
182. A. Hill, P. E. Ross and I. A. Bouchier, *Steroids*, 1981, **37**, 393.

CHAPTER 3

Bile-Acid Induction of Apoptosis in Relation to Gastrointestinal Cancer

KATERINA DVORAK,[a, b, d] HARRIS BERNSTEIN,[a, b]
CLAIRE M. PAYNE,[a, b] CAROL BERNSTEIN[a, d] AND
HARINDER GAREWAL[b, c, d]

[a] Department of Cell Biology and Anatomy, College of Medicine, University of Arizona, Tucson AZ 85724, USA; [b] Arizona Cancer Center and; [c] Department of Internal Medicine, University of Arizona, Tucson AZ 85724, USA; [d] Hematology/Oncology Southern Arizona Veterans Affairs Health Care System, Tucson, AZ 85723, USA

3.1 Introduction

Apoptosis is a physiological mode of cell death involving an orchestrated series of biochemical events. Apoptosis (from Greek *apo -, ptosis*-falling) is the normal cellular mechanism through which aged or damaged cells are eliminated from the body. The morphological changes defining apoptosis include chromatin condensation, margination of nuclear chromatin, cell shrinkage and nuclear fragmentation. Apoptotic cells form apoptotic bodies containing intact cytoplasmic organelles or fragments of the nucleus. These apoptotic bodies are eventually engulfed by resident phagocytic cells. Tumour cells may acquire resistance to apoptosis by the expression of anti-apoptotic proteins, or by the down-regulation or mutation of pro-apoptotic proteins.

Numerous studies (Table 3.1) indicate that exposure to high physiological levels of bile acids induces apoptosis in cells of the gastro-intestinal (GI) tract.

Issues in Toxicology
Bile Acids: Toxicology and Bioactivity
Edited by Gareth Jenkins and Laura J. Hardie
© Royal Society of Chemistry, 2008

Table 3.1   Bile acids induce apoptosis in cells of the GI tract.

Target organ/tissue/cell	Bile acid(s) that induced apoptosis	References
Esophagus (cultured human epithelial cells)	[a]GCDOC, GDOC, TC, TCDOC, TDOC	69
Esophagus (biopsies from normal human squamous epithelium)	DOC	46
Stomach (cultured human gastric epithelial cells)	DOC	70
Pancreas (mouse freshly isolated acinar cells)	Taurolithocholic acid 3-sulfate	71
Liver (cultured or freshly isolated hepatocytes)	DOC, GDOC, GCDOC, GC	72–81
Liver (tissue sections from rats fed bile acids)	DOC	81
Colon (cultured human tumour derived colonic epithelial cells)	DOC, CDOC	15,16,55,82–90
Colon (biopsies from normal human colonic mucosa)	DOC	1,27,28,52,91,92

[a]Abbreviations used for bile acids in Tables 3.1–3.6: DOC (deoxycholate), LC (lithocholate), CDOC (chenodeoxycholate), C (cholate), GDOC (glycodeoxycholate), TDOC (taurodeoxycholate), GCDOC (glycochenodeoxycholate), TCDOC (taurochenodeoxycholate), GC (glycocholate), TC (taurocholate).

These studies were generally carried out at bile-acid concentrations that occur naturally in humans (*e.g.* for colon cells, after a high-fat meal; for esophageal cells, in individuals with repeated heartburn due to duodeno-gastroesophageal reflux; or in cholestatic liver disorders). Taken together, these studies indicate that induction of apoptosis by bile acid is likely a frequent challenge for cells throughout the GI tract.

Other studies (Table 3.2) indicate that exposure to high physiological concentrations of bile acids, if repeated over a long period, increases the risk of GI cancer. A reasonable hypothesis is that bile acids act by a common underlying mechanism at various sites within the GI tract. Nevertheless, conditions vary widely from site to site within the GI tract, and it is certainly possible that at any particular site some factor(s) other than bile-acid exposure, or in combination with bile-acid exposure, is more important in carcinogenesis at that site.

The mechanism by which bile acids cause apoptosis, and then increase the risk for cancer has multiple steps. Induction of apoptosis by bile acids appears to be caused, at least in part, by oxidative/nitrosative stress and consequent DNA damage (Figure 3.1). Therefore, we summarise, below, evidence for the induction of oxidative/nitrosative stress and DNA damage by bile acids in cells of the GI tract, and their relationship to apoptosis. The cell's short-term response to oxidative/nitrosative stress and DNA damage is apparently to protect itself by adjusting gene expression to mitigate harmful effects. Some of these short-term immediate responses involve NF-κB activation (Figure 3.1), which can result in apoptosis inhibition. But, if these protective adjustments fail,

Table 3.2   Studies implicating bile acids in GI cancer.

Type of cancer	Evidence for association of bile acid(s) with cancer	References
Esophagus	Barrett's esophagus (BE) (metaplasia of the esophagus) is a major risk factor for esophageal adenocarcinoma development. Individuals with BE have increased duodeno-gastroesophageal reflux, and increased exposure to cytotoxic bile acids in their refluxate.	93,94
	Individuals with early esophageal adenocarcinoma have an even higher exposure to bile than individuals with uncomplicated BE.	95
	In rats, esophagogastroduodenal anastomosis, which induces reflux of gastric acid and bile, without concomitant chemical carcinogen treatment, can lead to esophageal adenocarcinoma.	96
	In rats, esophagojejunostomy with total gastrectomy to induce reflux of duodenal juice alone resulted in BE and esophageal adenocarcinoma in 16 weeks without any carcinogen.	97
Stomach	In the remnant stomach of rats after gastrectomy, bile acids are implicated in gastric cancer due to duodenogastric reflux.	98
	In humans, duodenogastric reflux is also implicated in gastric stump carcinoma.	99
	In rats, taurocholate enhances stomach tumourigenesis induced by N-methyl-N'-nitro-N-nitrosoguanidine.	100
Pancreas	Most adenocarcinomas of the pancreas occur in the head of the gland, which is in close proximity to bile.	101
	Bile reflux into the pancreatic ducts increased the number and incidence of intra-ductal papillary carcinomas in hamsters receiving weekly injections of N-nitrosobis(2-oxopropyl)amine. This is a model for human pancreaticobiliary maljunction, a congenital anomaly consisting of a union of the pancreatic and bile ducts, which causes reflux of bile into the pancreatic ducts and substantially increases the risk of pancreatic carcinoma.	102
Liver	A familial deficiency in the bile-acid export pump, which conveys bile acids from hepatocyte cytoplasm into bile canaliculi, increases the incidence of hepatocellular carcinoma in children.	103
Small intestine	In humans, about 57% of all adenocarcinomas of the small intestine occur in a 7-cm length of the duodenum that accounts for only about 1% of the entire length of the small intestine. Most of these adenocarcinomas occur in a small region around the Ampulla of Vater, where bile and pancreatic secretions enter the small intestine.	104

Table 3.2 (*Continued*).

	Persons with familial adenomatous polyposis are at increased risk for adenomas and cancers of the small and large intestines. In the small intestine, these lesions occur mainly around the Ampulla of Vater, where their distribution parallels mucosal exposure to bile.	105,106
	Administration of CDOC increased duodenal tumours near the Ampulla of Vater in a mouse model of familial adenomatous polyposis.	107
Colon	Epidemiological studies have found that fecal bile-acid concentrations are increased in populations with high incidence of colon cancer.	108–116
	In rats, LC, TDOC, DOC, C and CDOC have a promoting effect on colon carcinogenesis after intra-rectal instillation of N-methyl-N'-nitro-N-nitroso-guanidine or of N-methyl-N-nitrosourea.	117–120

proteins that signal apoptosis are then expressed and activated. However, repeated long-term exposure to bile acids appears to lead to natural selection of an epithelial cell population resistant to induction of apoptosis and prone to progression to cancer (Figure 3.1). Upon incurring further DNA damage resulting from continued exposure to bile acids and/or other genotoxins or low environmental pH, cells of the GI tract likely experience repeated replication errors, leading to mutation and consequently to increased risk of progression to cancer (Figure 3.1). Thus, we review, below, evidence that long-term exposure to bile acids selects for apoptosis resistance, that bile acids enhance the frequency of mutants, and that apoptosis resistance leads to increased mutation. In addition, we review recent emerging evidence that autophagy complements apoptosis in protecting against deleterious effects of bile acids. Autophagy is a major intra-cellular pathway for the degradation and recycling of long-lived proteins and cytoplasmic organelles. Autophagy is ordinarily a survival strategy for cleaning up damaged cellular components, especially in response to cellular stresses. However, when stresses on the cell are so great that autophagy is excessive, autophagy itself becomes cytotoxic and leads to autophagic cell death.

3.2 Bile Acids Induce Oxidative/Nitrosative Stress in Cells of the GI Tract

Reactive oxygen species (ROS) are a common mediator of apoptosis.[1] Induction of apoptosis by bile acids appears to be caused, at least in part, by oxidative stress and consequent DNA damage. Unrepaired DNA damage can trigger apoptosis.[2] Table 3.3 lists studies indicating that bile acids induce production of ROS and reactive nitrogen species (RNS) in cells of the GI tract. Table 3.4

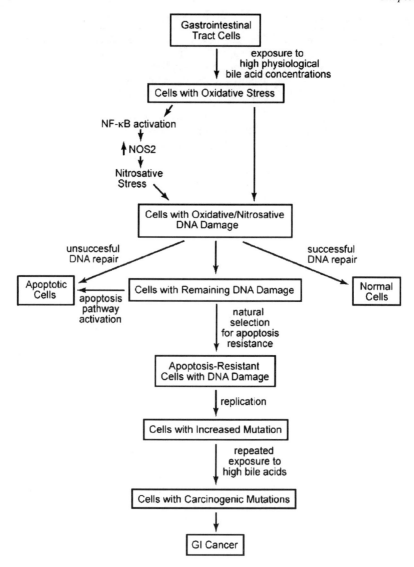

Figure 3.1 Pathway to GI cancer through bile-acid-induced oxidative DNA damage, leading to enhanced mutation.

summarises studies showing that bile acids cause DNA damage in these cells. Several groups reported that the type of DNA damage introduced by bile acids is oxidative DNA damage, suggesting that oxidative stress is a significant cause of the DNA damage.[3-9] These studies are listed in both Tables 3.3 and 3.4.

Bile acids likely generate ROS and RNS from multiple metabolic sources as previously reviewed.[10] One likely source of ROS is bile-acid-induced damage to mitochondria that causes the electron transport chain to "leak" electrons to O_2

Table 3.3   Bile acids induce oxidative/nitrosative stress in cells of the GI tract.

Target organ/tissue/cell	Bile acid(s)	Reference(s)
Esophagus (human tissues and cultured cells)	DOC or cocktail containing GC, TC, GDOC, GCDOC & DOC	7,8
Esophagus (mouse model of BE)	DOC (as dietary supplement; also zinc deficiency)	9
Stomach (cultured gastric cancer cells)	TCDOC	121
Biliary tract (immortalised rat cholangiocytes)	GCDOC	5
Liver (isolated rat hepatocytes)	GCDOC, TCDOC, GC	50,75,79,81,122,123
Liver (rat hepatocytes and cultured human hepatoma cells)	Taurolithocholate-3-sulfate	124
Colon (cultured human tumour derived colonic epithelial cells)	DOC, LC	3,4,35,85,121
Colon (mouse colonic mucosa)	DOC	6,125

Table 3.4   Bile acids induce DNA damage in cells of the GI tract.

Target organ/tissue/cell	Bile acid(s)	References
Esophagus (cultured human cells)	DOC or Cocktail containing GC, TC, GDOC, GCDOC & DOC at pH 4	7,8,17
Esophagus (biopsies from human Barrett's esophagus metaplastic tissue)	Cocktail containing GC, TC, GDOC, GCDOC & DOC	7
Esophagus (mouse model of esophagitis and BE)	DOC (as dietary supplement; also zinc deficiency)	9
Liver (human hepatoma cell line)	DOC	21
Biliary tract (immortalised mouse cholangiocytes)	GCDOC	5
Colon (cultured human tumour derived epithelial cells)	DOC, LC	3,4,14–16,19,22
Colon (freshly isolated colonic cells from human, mouse, rat)	LC	13,18
Colon (epithelial cells in rat tissue sections)	DOC (*ex-vivo* exposure)	20
Colon (epithelial cells from mouse model of colitis	DOC (as dietary supplement)	6

to form superoxide. For instance, mitochondrially derived ROS are induced by treatment with a mixture of bile acids and low pH.[7] Bile acids also can act at the cell membrane to release arachidonic acid, which is a substrate for cyclo-oxygenase and lipo-oxygenase. These enzymes partially reduce O_2 to super-oxide during synthesis of prostaglandins and leukotrienes, respectively.

Micromolar quantities of RNS are generated primarily by nitric oxide synthase 2 (NOS2), an enzyme that is up-regulated during colon-cancer progression.[11] As discussed below, deoxycholate (DOC), a hydrophobic secondary bile acid, activates the redox-sensitive transcription factor NF-κB, resulting in increased levels of NOS2 and enhanced S-nitrosylation of proteins.[12] Additional sources of bile-acid-induced ROS and RNS are also likely.[10]

3.3 Bile Acids Induce DNA Damage in Cells of the GI Tract

Table 3.4 lists studies indicating that bile acids induce DNA damage in cells of the GI tract. In these studies, DNA damage was measured or inferred by a variety of techniques. Direct measurements included the Comet assay that measures strand breaks by single-cell electrophoresis,[3,4,13–17] nucleoid sedimentation and alkaline elution,[18] and 8-OHdG assay[5–7,9]). Indirect measurements include increased expression/activation of the DNA-repair proteins BRCA1 and PARP[19,20]), increased expression of the DNA-damage-responsive genes GADD34, GADD45 and GADD153,[21,22] and increased frequency of micronuclei.[8]

3.4 Bile Acids Activate NF-κB as a Stress Response and Survival Pathway Throughout the GI Tract

Hydrophobic bile acids are a group of endogenous compounds that are necessary for the digestion of fat (through their detergent action), but whose persistently increased levels can induce inflammation and promote cancer. After the ingestion of fat, increased levels of bile acids are released into the duodenum from the gallbladder. DOC is one of the most abundant and cytotoxic hydrophobic bile acids in the gut. We have shown that DOC induces the activation of NF-κB in cultured colonic epithelial cells (*e.g.* HCT-116, HT-29)[20,23] and hepatoma cells.[21] Deoxycholate-induced NF-κB activation has also been documented in esophageal cells.[24,25] These *in-vitro* findings indicate the importance of these dietary-related detergents in the activation of proinflammatory pathways. Our group has recently developed two unique dietary-related mouse models of inflammation in the lower and upper GI tract.[6,9] In the first model, high physiologic levels of DOC in the diet induced colitis, which was associated with increased mRNA expression of Pellino 1, a major inflammatory cytokine required for the activation of NF-κB in response to IL-1.[6] In the second model, high physiologic levels of DOC in a zinc-deficient diet resulted in the induction of esophagitis and the development of a Barrett's esophagus-like lesion.[9] Using an immunohistochemical approach, NF-κB was shown to be present in the nucleus of epithelial cells in the areas of inflammation, providing *in-vivo* confirmation of the *in-vitro* findings.

What are the potential mechanisms by which DOC can activate NF-κB? In a study of over 40 inhibitors of many signal-transduction pathways, we have found that DOC can activate NF-κB through multiple mechanisms involving NAD(P)H oxidase, Na^+/K^+-ATPase, Cytochrome P450, Ca^{++} ions and the terminal mitochondrial respiratory complex IV.[26] A potential outcome of persistent stress is the induction of apoptosis, which will eliminate damaged cells from the population. It is becoming evident, however, that clones of apoptosis-resistant cells then proliferate, thereby contributing to field defects of apoptosis resistance.[27-29] Proof of this principle was provided by exposing apoptosis-sensitive cells for months to increasing concentrations of DOC. This resulted in the development of apoptosis-resistant cells,[30] which were associated with an increase in the p50 and p65 sub-units of NF-κB, and the increased activation of NF-κB assessed using a monoclonal antibody against an epitope of the p65 sub-unit normally masked by IκB. Anti-sense oligonucleotides directed at the p65 sub-unit of NF-κB resulted in sensitisation to apoptosis, providing *in-vitro* evidence that NF-κB contributes to cell survival in these apoptosis-resistant colonic cancer cells.[30]

NF-κB is a redox-sensitive transcription factor that functions in stress-response pathways of the GI tract in response to various stressors such as microbes [*e.g.* enteroinvasive *Escherichia coli, Helicobacter pylori*[31,32]], low extra-cellular pH[32-34]), tobacco components [*e.g.* nicotine[35]], inflammatory molecules [*e.g.* TNF-α[36]] and DNA-damaging agents, with double-strand breaks being the most likely candidate among all possible types of DNA damages.[37] Persistent activation of NF-κB can further lead to the induction of inflammatory molecules, such as NO, IL-1α, IL-6 and IL-8 resulting in chronic inflammation, a predisposing condition for tumourigenesis throughout the GI tract.[38] NF-κB is constitutively elevated in laryngopharyngeal cancer, esophageal adenocarcinoma, colonic adenocarcinoma and pancreatic cancer, where it serves to activate anti-apoptotic pathways. Activated NF-κB is also found in pre-cancerous lesions of the GI tract (*e.g.* Barrett's esophagus), where it represents a critical event in the neoplastic progression associated with Barrett's esophagus.[39]

The importance of NF-κB to inflammation, apoptosis resistance and tumour progression has resulted in the development of unique NF-κB inhibitors as part of cancer therapeutic regimens for GI and other cancers.[40-42] Efforts are also being made to understand the efficacy of using natural substances obtained from plants, such as feverfew (*e.g.* parthenolide), bee "glue" (*e.g.* caffeic acid phenylethyl ester), tea (*e.g.* EGCG), spices (*e.g.* curcumin from turmeric) and mulberry figs (*e.g.* morin, a flavone) for the prevention both of persistent NF-κB activation and of the development of inflammatory pre-neoplastic lesions.

3.5 Bile Acids Increase Mutant Cells in the GI Tract

The observation that bile acids cause DNA damage (Table 3.4) suggests that bile acids should increase the frequency of mutation since unrepaired DNA damage causes replication errors. Table 3.5 lists the studies showing that bile acids cause an increase in mutant cells in the GI tract. *In vitro*, DOC treatment

Table 3.5 Bile acids increase the incidence of mutant cells in the GI tract.

Target organ/tissue/cell	Bile acid(s)	Evidence for induced mutation	References
Esophagus (squamous carcinoma KYSE-30 cells)	DOC	Increase in *p53* mutations	8
Esophagus (Big Blue F1 *lacI* transgenic rats)	Increased duo-deno-gastric-esophageal reflux	Significant increase in *lacI* mutations	43
Colon (rat tumour cells)	DOC	Increase in *K-ras* point mutations in colon tumours	44

of esophageal cells increased the frequency of cells with a GC to AT mutation in the p53 gene.[8] *In vivo,* Theisen *et al.* used a rat model for measuring mutant cells. Big Blue F1 *lacI* transgenic rats were subjected to esophagoduodenostomy to surgically create duodeno-gastric-esophageal reflux.[43] The frequency of mutant *lacI* cells in the esophageal mucosa of surgically altered animals was significantly higher than of nonoperated animals. They found that 46% of the mutant cells were altered at CpG dinucleotide sites and that the majority (61%) were C to T or G to A transitions. They noted that this pattern is similar to the pattern seen in human esophageal adenocarcinoma, suggesting that duodenoesophageal reflux is carcinogenic. In addition, they pointed out that, in general, tumours with the highest frequency of mutant cells with CpG transition mutations in the p53 gene arise in those parts of the GI tract where exposure to bile occurs. In a further *in-vivo* study, the azoxymethane-treated rat model of colon tumourigenesis, DOC increased the incidence of colon tumours as well as the incidence of cells with K-ras point mutations in the tumours.[44] If bile acids increase the frequency of mutant cells, as the data indicate, bile-acid exposure either induces mutations directly, or allows for the clonal outgrowth of mutant cells. That bile acids can induce mutations directly is indicated by the finding in a sensitive bacterial fluctuation test, based on the Ames Salmonella mutagen test system, that chenodeoxycholic acid is significantly mutagenic.[45] Some of the mutations caused by bile acids would be expected, by chance, to cause aberrant expression of oncogenes and tumour-suppressor genes leading to cancer (Figure 3.1).

3.6 Repeated Exposure to High Concentrations of Bile Acids Probably Selects for Apoptosis Resistance in Cells of the GI Tract

A number of observations (Table 3.6) indicate that tumours arise in an apoptosis-resistant pre-malignant cell population. Barrett's esophagus (BE) is a pre-malignant lesion of the distal esophagus in which squamous epithelial cells

Table 3.6 Observations indicating that development of apoptosis resistance often occurs early in cancer progression in the GI tract.

Target organ/tissue/cell	Evidence for apoptosis resistance in pre-malignant tissue	References
Esophagus (cells of BE, a pre-neoplastic lesion)	Increased resistance to DOC-induced apoptosis in BE cells compared to control cells. Increased expression of anti-apoptotic proteins Bcl-x_L, and Mcl-1	47,126
Bile-duct epithelial cells (in pre-neoplastic bile-duct inflammatory disease)	Strongly increased expression of anti-apoptotic protein Mcl-1	49
Pancreas (pre-malignant ductal lesions)	In a mouse model of pancreatic cancer there is increased expression of the anti-apoptotic protein Bcl-x_L and decreased expression of pro-apoptotic Bax protein	51
Colon [biopsies from colonic mucosa (nontumour tissue) of patients with colon cancer]	Increased resistance to DOC-induced apoptosis of colonic epithelial cells of cancer patients or increased expression of anti-apoptotic protein Bcl-x_L.	1,27,28,52,127

are replaced by metaplastic intestinal-like columnar epithelium that contains goblet cells. BE cells are resistant to apoptosis induced by DOC compared to cells of the esophageal squamous epithelium and normal colon epithelium[46] (Table 3.6). Anti-apoptotic proteins Bcl-x_L, and Mcl-1 are expressed at higher levels in BE tissue than in nearby epithelial tissue, suggesting that these proteins may be contributing to the observed apoptosis resistance of BE.[47] Furthermore, it was suggested that alterations in apoptotic balance lead to the transformation from BE to adenocarcinoma, since increased expression of anti-apoptotic Bcl-x_L and decreased expression of pro-apoptotic Bax are hallmarks for progression to adenocarcinoma.[48] Patients with BE have frequent reflux episodes associated with increased exposure to bile acids (Table 3.6) suggesting that this exposure may select for BE cells resistant to bile-acid-induced apoptosis, and/or induce the metaplastic process associated with the conversion of squamous epithelium to columnar epithelium with mucin granules (*i.e.* BE).

In bile-duct epithelial cells, the anti-apoptotic protein Mcl-1 is strongly expressed in the pre-neoplastic bile-duct inflammatory disease, primary sclerosing cholangitis.[49] This condition appears to involve intra-cellular accumulation of hydrophobic bile acids.[50]

In a transgenic mouse model of ductal pancreatic adenocarcinoma, the development of ductal carcinoma is preceded by trans-differentiation of acinar cells to ductal-like cells.[51] The anti-apoptotic protein Bcl-x_L is highly expressed in pre-malignant tubular complexes formed of ductal-like cells compared to acinar cells of normal appearing areas. The pro-apoptotic protein Bax shows the opposite pattern of expression.[51] These observations indicate an imbalanced production of pro-apoptotic and anti-apoptotic proteins in the direction of apoptosis resistance in the pre-malignant cells.

In humans, epithelial cells of the flat non-neoplastic colonic mucosa of individuals with colon cancer often have reduced capacity to undergo DOC-induced apoptosis[1,27,28,52] (Table 3.6) suggesting that areas of apoptosis-resistant epithelium may give rise to colon cancer. Individuals with colon cancer, on average, have experienced elevated levels of bile-acid exposure (Table 3.2) suggesting that exposure to high levels of bile acids may select for cells resistant to induction of apoptosis. This idea is supported by the finding, in rats, that repeated feeding of a diet containing 0.2% cholic acid resulted in the development of increased resistance to apoptosis of colon crypt cells, both in aberrant crypt foci and in normal crypts.[53] Apoptosis-resistant crypts may then spread through the flat mucosa by a process of crypt fission.[54]

Repeated long-term exposure of cultured human colonic epithelial cell lines to sub-lethal concentrations of DOC allowed selection for cell lines that were resistant to DOC-induced apoptosis,[30] suggesting that similar selection by repeated DOC exposure may occur in the colon. A microarray analysis of three apoptosis-resistant cell lines indicated that changes had occurred in expression at the mRNA level of numerous genes that may play a role in apoptosis and early carcinogenesis.[30] Apoptosis resistance in these cell lines was due, in part, to the nitric oxide synthase/nitric oxide/guanylate cyclase/cyclic guanosine monophosphate/cGMP-activated protein kinase signalling module.[55] As evidence that this NO-signalling module contributes to apoptosis resistance, cells that were pre-treated with pharmacological inhibitors/antagonists to different members of the module became sensitised to apoptosis. A proteomic analysis of these apoptosis-resistant cell lines indicated that numerous anti-apoptotic proteins were over-expressed (*e.g.* 5-LOX, PI3K, P-cadherin and cyclin D3), and that numerous pro-apoptotic proteins were under-expressed (*e.g.* Bid, CaM kinase II and JNK-1).[56]

3.7 Apoptosis Resistance and Increased Mutation

Although evidence is not yet available for bile-acid-exposed cells of the GI tract, evidence from other cell types suggests that reduced capability of DNA-damaged cells to undergo apoptosis leads to increased replication errors and consequent mutation. Suppression of apoptosis by over-expression of Bcl-2 or Bcl-x_L in human lymphoblastoid cell lines markedly elevated the level of mutation induced by the oxidising agent hydrogen peroxide,[57] as well as mutation induced by ionising radiation.[58] Also, Bcl-2 over-expression caused an increase in spontaneous mutation as well as mutagenesis by UV and ionising radiation, as determined by a fluctuation analysis).[59] Furthermore, suppression of apoptosis by Bcl-2-enhanced oxidative DNA damage and mutagenesis induced by 1,4-hydroquinone- and 1,4-benzoquinone in HL-60 leukemia cells.[60] When mice were exposed *in vivo* to repeated ionising radiation, there was apparent selection for reduced apoptosis in lymphoid cells leading to increased deletion mutation.[61] These findings suggest that apoptosis-resistant cells may be more likely to evolve to a malignant state because of their higher mutation rate as well as the growth advantage provided by being less prone to die when damaged, as indicated in Figure 3.1.

3.8 Modes of Cell Death: Apoptosis, Autophagy and Necrosis

Epithelial cells of the GI tract are exposed to various agents (as indicated in previous sections) and conditions that activate stress-response pathways. This results in either cell survival or cell death. Those cells with inadequate cellular defenses may die by one of three major cell death mechanisms, apoptosis, autophagy or necrosis,[62] depending upon the particular stress conditions.

Apoptosis and autophagy do not generally result in inflammation since the intra-cellular contents are not released into the extra-cellular microenvironment, and the dying cells are eliminated through a process of rapid cellular shrinkage (apoptosis) or a process of cellular atrophy as a result of auto-digestion (autophagy) followed by engulfment by macrophages.[63,64] Thus, the catastrophic type of cell death that is avoided *in vivo* is necrosis, a process of cellular swelling and lysis that results in the release of the intra-cellular contents into the microenvironment. One of the intra-cellular proteins that is released during the necrotic process is HMGB1 (high-mobility group box 1), which has the ability to trigger inflammation.[65] It is well known that chronic inflammation of the GI tract can lead to cancer. Therefore, avoidance of a mode of cell death that induces inflammation is desirable.

Our research group has been specifically studying the role of bile acids in inducing cellular stresses and cell death in both the upper and lower GI tract. We hypothesise that prolonged periods of bile-acid-induced cell stress leading to repeated apoptotic episodes within a population of colon epithelial cells will result in the clonal selection of apoptosis-resistant cells.[27] Apoptosis-resistant cells are prone to mutation, ultimately resulting in neoplastic clones that lead to cancer (see sections on apoptosis resistance and increased mutation, above). Many chemotherapeutic regimens now include an agent targetted to the apoptosis-resistant pathway, to restore apoptosis competence, resulting in more effective killing of malignant cells.

Autophagy can be an important survival mechanism at lower levels of cell damage, by degrading and recycling damaged proteins and organelles.[62] On the other hand, since autophagy is a process of self-digestion, excessive activation of autophagy by high levels of cell damage or by the autophagy inducer rapamycin, for example, may tip the balance in favor of autophagic cell death rather than survival.[66]

Although autophagy is biochemically and morphologically distinct from apoptosis, there is significant cross-talk between these signalling pathways.[67,68] It is, therefore, not surprising that bile acids that activate the apoptotic pathway and can generate apoptosis resistance in colon epithelial cells,[30] may also activate autophagy and modulate the expression of various genes associated with the autophagic signalling pathway. We showed that the apoptosis-resistant HCT-116 cell lines that were developed in our laboratory after persistent exposure to increasing concentrations of deoxycholate[30] undergo activation of lysosomes and modulation of genes associated with the autophagic pathway. These autophagy-related genes that exhibit increased expression include six *rab* genes involved in vesicle transport, PI(3)K, 2 sub-units of the lysosomal proton (H^+)-translocating ATPase, cathepsin D, Lamp-1 (lysosomal-associated membrane protein 1), and prenylcysteine lyase, a lysosomal enzyme involved in the degradation of prenylated proteins. In addition, we reported that chronic feeding of wild-type B6.129 mice with deoxycholate (an apoptosis-inducing agent) added to the diet, results in an increase in APG4, a cysteine protease that acts during the formation of auto-phagosomes, and Sh3kbp1, an endogenous inhibitor of PI(3)K, a negative regulator of auto-phagy.[6]

We have recently shown, in Seg-1 cells derived from a patient with esophageal adenocarcinoma, that bile acids and low pH induce the dramatic formation of autophagic vacuoles (unpublished data). We also showed that persistent treatment of Het-1A cells (derived from normal human esophageal squamous epithelium) with a bile-acid cocktail and low pH resulted in the development of mutated clones that contain a partly duplicated chromosome 16 (unpublished data from a CGH microarray). In addition to other protective genes, this chromosome 16 duplication resulted in the duplication of the major ATPase necessary for the acidification of vesicles associated with the autophagic process.

As reviewed above, apoptosis resistance often occurs early in cancer progression in the GI tract, and is also often present in cancers. Dietary agents that enhance autophagy may be useful in chemoprevention by increasing autophagic cell death of pre-malignant apoptosis-resistant cells. Increased autophagy may similarly be therapeutic by increasing autophagic cell death of cancer cells.

3.9   Overview

The findings reviewed here indicate that bile acids play a key etiologic role in GI cancer. Since bile acids at high physiological levels induce apoptosis, we have proposed that repeated high exposure over decades leads to selection for apoptosis-resistant cells. We have reviewed evidence that bile acids also cause oxidative stress, DNA damage and mutation. Thus, cells that survive bile-acid exposure, often because they are apoptosis resistant, will be subject to increased mutation leading to cancer. We consider that, in humans as distinct from short-term rodent models, bile acids can act as carcinogens when there is repeated exposure to high levels over several decades.[10]

Acknowledgments

This work was supported in part by NIH Institutional Core Grant CA23074, NIH PPG CA72008, NIH 1R21CA111513-01A1, NIH 5 RO1 CA119087, NIH(NCI) SPORE Grant 1 P50 CA95060, Arizona Disease Control Research Commission Grants (#10016, #6002), Arizona Biomedical Research Commission Grants (#0012 & #0803), VAH Merit Review Grant 2HG, and Biomedical Diagnostics & Research, Inc., Tucson, Arizona.

References

1. C. M. Payne, C. Bernstein and H. Bernstein, Apoptosis overview emphasizing the role of oxidative stress, DNA damage and signal-transduction pathways, *Leuk. Lymphoma*, 1995, **19**(1–2), 43.
2. C. Bernstein, H. Bernstein, C. M. Payne and H. Garewal, DNA repair/pro-apoptotic dual-role proteins in five major DNA repair pathways: failsafe protection against carcinogenesis, *Mutat. Res.*, 2002, **511**(2), 145.
3. L. A. Booth, I. T. Gilmore and R. F. Bilton, Secondary bile-acid induced DNA damage in HT29 cells: are free radicals involved? *Free Radic. Res.*, 1997, **26**(2), 135.
4. M. Venturi, R. J. Hambly, B. Glinghammar, J. J. Rafter and I. R. Rowland, Genotoxic activity in human faecal water and the role of bile acids: a study using the alkaline comet assay, *Carcinogenesis*, 1997, **18**(12), 2353.
5. D. Komichi, S. Tazuma, T. Nishioka, H. Hyogo and K. Chayama, Glycochenodeoxycholate plays a carcinogenic role in immortalized mouse cholangiocytes via oxidative DNA damage, *Free Radic. Biol. Med.*, 2005, **39**(11), 1418.
6. H. Bernstein, H. Holubec, C. Bernstein, N. Ignatenko, E. Gerner, K. Dvorak, D. Besselsen, L. Ramsey, M. Dall'Agnol, K. A. Blohm-Mangone, J. Padilla-Torres, H. Cui, H. Garewal and C. M. Payne, Unique dietary-related mouse model of colitis, *Inflamm. Bowel Dis.*, 2006, **12**(4), 278.
7. K. Dvorak, C. M. Payne, M. Chavarria, L. Ramsey, B. Dvorakova, H. Bernstein, H. Holubec, R. E. Sampliner, N. Guy, A. Condon, C. Bernstein, S. B. Green, A. Prasad and H. S. Garewal, Bile acids in combination with low pH induce oxidative stress and oxidative DNA damage: relevance to the pathogenesis of Barrett's oesophagus, *Gut*, 2007, **56**(6), 763.
8. G. J. Jenkins, F. R. D'souza, S. H. Suzen, Z. S. Eltahir, S. A. James, J. M. Parry, P. A. Griffiths and J. N. Baxter, Deoxycholic acid at neutral and acid pH, is genotoxic to oesophageal cells through the induction of ROS: the potential role of anti-oxidants in Barrett's oesophagus, *Carcinogenesis*, 2007, **28**(1), 136.
9. N. C. Guy, H. Garewal, H. Holubec, H. Bernstein, C. M. Payne, C. Bernstein, A. Bhattacharyya and K. Dvorak, A novel dietary-related

model of esophagitis and Barrett's esophagus, a premalignant lesion, *Nutrition and Cancer*, 2007, **59**(2), 217.

10. H. Bernstein, C. Bernstein, C. M. Payne, K. Dvorakova and H. Garewal, Bile acids as carcinogens in human gastro-intestinal cancers, *Mutat. Res.*, 2005, **589**(1), 47.

11. C. M. Payne, C. Bernstein, H. Bernstein, E. W. Gerner and H. Garewal, Reactive nitrogen species in colon carcinogenesis, *Antioxid. Redox. Signal.*, 1999, **1**(4), 449.

12. M. Dall'Agnol, C. Bernstein, H. Bernstein, H. Garewal and C. M. Payne, Identification of S-nitrosylated proteins after chronic exposure of colon epithelial cells to deoxycholate, *Proteomics*, 2006, **6**(5), 1654.

13. B. L. Pool-Zobel and U. Leucht, Induction of DNA damage by risk factors of colon cancer in human colon cells derived from biopsies, *Mutat. Res.*, 1997, **375**(2), 105.

14. L. A. Booth and R. F. Bilton, Genotoxic potential of the secondary bile acids: a role for reactive oxygen species, in *DNA and Free Radicals: Techniques, Mechanisms & Applications*, O.I. Arouma and B. Halliwell, Editor, 1998, OICA International: London, 161.

15. A. Powolny, J. Xu and G. Loo, Deoxycholate induces DNA damage and apoptosis in human colon epithelial cells expressing either mutant or wild-type p53, *Int. J. Biochem. Cell Biol.*, 2001, **33**(2), 193.

16. B. Glinghammar, H. Inoue and J. J. Rafter, Deoxycholic acid causes DNA damage in colonic cells with subsequent induction of caspases, COX-2 promoter activity and the transcription factors NF-kB and AP-1, *Carcinogenesis*, 2002, **23**(5), 839.

17. A. J. Jolly, C. P. Wild and L. J. Hardie, Acid and bile salts induce DNA damage in human oesophageal cell lines, *Mutagenesis*, 2004, **19**(4), 319.

18. M. S. Kulkarni and K. L. Yielding, DNA damage and repair in epithelial (mucous) cells and crypt cells from isolated colon, *Chem. Biol. Interact.*, 1985, **52**(3), 311.

19. D. F. Romagnolo, R. B. Chirnomas, J. Ku, B. D. Jeffy, C. M. Payne, H. Holubec, L. Ramsey, H. Bernstein, C. Bernstein, K. Kunke, A. Bhattacharyya, J. Warneke and H. Garewal, Deoxycholate, an endogenous tumour promoter and DNA damaging agent, modulates BRCA-1 expression in apoptosis-sensitive epithelial cells: loss of BRCA-1 expression in colonic adenocarcinomas, *Nutr. Cancer*, 2003, **46**(1), 82.

20. C. M. Payne, C. Crowley, D. Washo-Stultz, M. Briehl, H. Bernstein, C. Bernstein, S. Beard, H. Holubec and J. Warneke, The stress-response proteins poly(ADP-ribose) polymerase and NF-kappaB protect against bile salt-induced apoptosis, *Cell Death Differ.*, 1998, **5**(7), 623.

21. H. Bernstein, C. M. Payne, C. Bernstein, J. Schneider, S. E. Beard and C. L. Crowley, Activation of the promoters of genes associated with DNA damage, oxidative stress, ER stress and protein malfolding by the bile salt, deoxycholate, *Toxicol. Lett.*, 1999, **108**(1), 37.

22. D. W. Scott, S. Mutamba, R. G. Hopkins and G. Loo, Increased GADD gene expression in human colon epithelial cells exposed to deoxycholate, *J. Cell. Physiol.*, 2005, **202**(1), 295.

23. D. Washo-Stultz, C. L. Crowley-Weber, K. Dvorakova, C. Bernstein, H. Bernstein, K. Kunke, C. N. Waltmire, H. Garewal and C. M. Payne, Role of mitochondrial complexes I and II, reactive oxygen species and arachidonic acid metabolism in deoxycholate-induced apoptosis, *Cancer Lett.*, 2002, **177**(2), 129.

24. G. J. Jenkins, K. Harries, S. H. Doak, A. Wilmes, A. P. Griffiths, J. N. Baxter and J. M. Parry, The bile acid deoxycholic acid (DCA) at neutral pH activates NF-{kappa}B and induces IL-8 expression in oesophageal cells *in vitro*, *Carcinogenesis*, 2004, **25**(3), 317.

25. P. R. Debruyne, M. Witek, L. Gong, R. Birbe, I. Chervoneva, T. Jin, C. Domon-Cell, J. P. Palazzo, J. N. Freund, P. Li, G. M. Pitari, S. Schulz and S. A. Waldman, Bile acids induce ectopic expression of intestinal guanylyl cyclase C through nuclear factor-kappaB and Cdx2 in human esophageal cells, *Gastroenterology*, 2006, **130**(4), 1191.

26. C. M. Payne, C. Weber, C. Crowley-Skillicorn, K. Dvorak, H. Bernstein, C. Bernstein, H. Holubec, B. Dvorakova and H. Garewal, Deoxycholate induces mitochondrial oxidative stress and activates NF-kappaB through multiple mechanisms in HCT-116 colon epithelial cells, *Carcinogenesis*, 2007, **28**(1), 215.

27. C. Bernstein, H. Bernstein, H. Garewal, P. Dinning, R. Jabi, R. E. Sampliner, M. K. McCuskey, M. Panda, D. J. Roe, L. L'Heureux and C. Payne, A bile-acid-induced apoptosis assay for colon cancer risk and associated quality control studies, *Cancer Res.*, 1999, **59**(10), 2353.

28. H. Bernstein, H. Holubec, J. A. Warneke, H. Garewal, D. L. Earnest, C. M. Payne, D. J. Roe, H. Cui, E. L. Jacobson and C. Bernstein, Patchy field defects of apoptosis resistance and dedifferentiation in flat mucosa of colon resections from colon cancer patients, *Ann. Surg. Oncol.*, 2002, **9**(5), 505.

29. C. M. Payne, C. Bernstein and H. Bernstein, Field change of apoptosis resistance in colonic mucosa of patients with colorectal carcinoma, *J. Clin. Path.*, 2007, Electronic letters published (5 February, 2007).

30. C. L. Crowley-Weber, C. M. Payne, M. Gleason-Guzman, G. S. Watts, B. Futscher, C. N. Waltmire, C. Crowley, K. Dvorakova, C. Bernstein, M. Craven, H. Garewal and H. Bernstein, Development and molecular characterization of HCT-116 cell lines resistant to the tumour promoter and multiple stress-inducer, deoxycholate, *Carcinogenesis*, 2002, **23**(12), 2063.

31. K. La Ferla, D. Seegert and S. Schreiber, Activation of NF-kappaB in intestinal epithelial cells by *E. coli* strains isolated from the colonic mucosa of IBD patients, *Int. J. Colorectal Dis.*, 2004, **19**(4), 334.

32. D. O'Toole, M. M. Abdel-Latif, A. Long, H. J. Windle, A. M. Murphy, A. Bowie, L. A. O'Neill, D. G. Weir and D. Kelleher, Low pH and Helicobacter pylori increase nuclear factor kappa B binding in gastric

epithelial cells: a common pathway for epithelial cell injury? *J. Cell. Biochem.*, 2005, **96**(3), 589.

33. M. M. Abdel-Latif, J. O'Riordan, H. J. Windle, E. Carton, N. Ravi, D. Kelleher and J. V. Reynolds, NF-kappaB activation in esophageal adeno-carcinoma: relationship to Barrett's metaplasia, survival, and response to neoadjuvant chemoradiotherapy, *Ann. Surg.*, 2004, **239**(4), 491.

34. S. P. Duggan, W. M. Gallagher, E. J. Fox, M. M. Abdel-Latif, J. V. Reynolds and D. Kelleher, Low pH induces co-ordinate regulation of gene expression in oesophageal cells, *Carcinogenesis*, 2006, **27**(2), 319.

35. C. L. Crowley-Weber, K. Dvorakova, C. Crowley, H. Bernstein, C. Bernstein, H. Garewal and C. M. Payne, Nicotine increases oxidative stress, activates NF-kappaB and GRP78, induces apoptosis and sensitizes cells to genotoxic/xenobiotic stresses by a multiple stress inducer, deoxycholate: relevance to colon carcinogenesis, *Chem. Biol. Interact.*, 2003, **145**(1), 53.

36. N. L. Rhodus, B. Cheng, S. Myers, L. Miller, V. Ho and F. Ondrey, The feasibility of monitoring NF-kappaB associated cytokines: TNF-alpha, IL-1alpha, IL-6, and IL-8 in whole saliva for the malignant transformation of oral lichen planus, *Mol. Carcinog.*, 2005, **44**(2), 77.

37. S. Janssens and J. Tschopp, Signals from within: the DNA-damage-induced NF-kappaB response, *Cell. Death Differ.*, 2006, **13**(5), 773.

38. C. J. van der Woude, J. H. Kleibeuker, P. L. Jansen and H. Moshage, Chronic inflammation, apoptosis and (pre-)malignant lesions in the gastro-intestinal tract, *Apoptosis*, 2004, **9**(2), 123.

39. P. C. Konturek, A. Nikiforuk, J. Kania, M. Raithel, E. G. Hahn and S. Muhldorfer, Activation of NFkappaB represents the central event in the neoplastic progression associated with Barrett's esophagus: a possible link to the inflammation and over-expression of COX-2, PPARgamma and growth factors, *Dig. Dis. Sci.*, 2004, **49**(7–8), 1075.

40. M. Karin, NF-kappaB and cancer: mechanisms and targets, *Mol. Carcinog.*, 2006, **45**(6), 355.

41. H. J. Kim, N. Hawke and A. S. Baldwin, NF-kappaB and IKK as therapeutic targets in cancer, *Cell Death Differ.*, 2006, **13**(5), 738.

42. C. Van Waes, Nuclear factor-kappaB in development, prevention, and therapy of cancer, *Clin. Cancer Res.*, 2007, **13**(4), 1076.

43. J. Theisen, J. H. Peters, M. Fein, M. Hughes, J. A. Hagen, S. R. Demeester, T. R. Demeester and P. W. Laird, The mutagenic potential of duodenoesophageal reflux, *Ann. Surg.*, 2005, **241**(1), 63.

44. H. Narahara, M. Tatsuta, H. Iishi, M. Baba, N. Uedo, N. Sakai, H. Yano and S. Ishiguro, K-ras point mutation is associated with enhancement by deoxycholic acid of colon carcinogenesis induced by azoxymethane, but not with its attenuation by all-trans-retinoic acid, *Int. J. Cancer*, 2000, **88**(2), 157.

45. J. Watabe and H. Bernstein, The mutagenicity of bile acids using a fluctuation test, *Mutat. Res.*, 1985, **158**(1–2), 45.

46. K. Dvorakova, C. M. Payne, L. Ramsey, H. Bernstein, H. Holubec, M. Chavarria, C. Bernstein, R. E. Sampliner, C. Riley, A. Prasad and H. Garewal, Apoptosis resistance in Barrett's esophagus: *ex vivo* bioassay of live stressed tissues, *Am. J. Gastroenterol.*, 2005, **100**(2), 424.

47. K. Dvorakova, C. M. Payne, L. Ramsey, H. Holubec, R. Sampliner, J. Dominguez, B. Dvorak, H. Bernstein, C. Bernstein, A. Prasad, R. Fass, H. Cui and H. Garewal, Increased expression and secretion of interleukin-6 in patients with Barrett's esophagus, *Clin. Cancer Res.*, 2004, **10**(6), 2020.

48. C. J. van der Woude, P. L. Jansen, A. T. Tiebosch, A. Beuving, M. Homan, J. H. Kleibeuker and H. Moshage, Expression of apoptosis-related proteins in Barrett's metaplasia-dysplasia-carcinoma sequence: a switch to a more resistant phenotype, *Hum. Pathol.*, 2002, **33**(7), 686.

49. S. Kobayashi, N. W. Werneburg, S. F. Bronk, S. H. Kaufmann and G. J. Gores, Interleukin-6 contributes to Mcl-1 up-regulation and TRAIL resistance via an Akt-signalling pathway in cholangiocarcinoma cells, *Gastroenterology*, 2005, **128**(7), 2054.

50. R. J. Sokol, J. M. McKim, Jr., M. C. Goff, S. Z. Ruyle, M. W. Devereaux, D. Han, L. Packer and G. Everson, Vitamin E reduces oxidant injury to mitochondria and the hepatotoxicity of taurochenodeoxycholic acid in the rat, *Gastroenterology*, 1998, **114**(1), 164.

51. F. R. Greten, C. K. Weber, T. F. Greten, G. Schneider, M. Wagner, G. Adler and R. M. Schmid, Stat3 and NF-kappaB activation prevents apoptosis in pancreatic carcinogenesis, *Gastroenterology*, 2002, **123**(6), 2052.

52. H. Garewal, H. Bernstein, C. Bernstein, R. Sampliner and C. Payne, Reduced bile-acid-induced apoptosis in "normal" colorectal mucosa: a potential biological marker for cancer risk, *Cancer Res.*, 1996, **56**(7), 1480.

53. B. A. Magnuson, N. Shirtliff and R. P. Bird, Resistance of aberrant crypt foci to apoptosis induced by azoxymethane in rats chronically fed cholic acid, *Carcinogenesis*, 1994, **15**(7), 1459.

54. S. B. Garcia, H. S. Park, M. Novelli and N. A. Wright, Field cancerization, clonality and epithelial stem cells: the spread of mutated clones in epithelial sheets, *J. Pathol.*, 1999, **187**(1), 61.

55. C. M. Payne, C. N. Waltmire, C. Crowley, C. L. Crowley-Weber, K. Dvorakova, H. Bernstein, C. Bernstein, H. Holubec and H. Garewal, Caspase-6 mediated cleavage of guanylate cyclase alpha 1 during deoxycholate-induced apoptosis: protective role of the nitric oxide signalling module, *Cell Biol. Toxicol.*, 2003, **19**(6), 373.

56. H. Bernstein, C. M. Payne, K. Kunke, C. L. Crowley-Weber, C. N. Waltmire, K. Dvorakova, H. Holubec, C. Bernstein, R. R. Vaillancourt, D. A. Raynes, V. Guerriero and H. Garewal, A proteomic study of resistance to deoxycholate-induced apoptosis, *Carcinogenesis*, 2004, **25**(5), 681.

57. C. Cherbonnel-Lasserre and M. K. Dosanjh, Suppression of apoptosis by over-expression of Bcl-2 or Bcl-xL promotes survival and mutagenesis after oxidative damage, *Biochimie*, 1997, **79**(9-10), 613.

58. C. Cherbonnel-Lasserre, S. Gauny and A. Kronenberg, Suppression of apoptosis by Bcl-2 or Bcl-xL promotes susceptibility to mutagenesis, *Oncogene*, 1996, **13**(7), 1489.
59. Y. Saintigny, A. Dumay, S. Lambert and B. S. Lopez, A novel role for the Bcl-2 protein family: specific suppression of the RAD51 recombination pathway, *Embo J.*, 2001, **20**(10), 2596.
60. M. L. Kuo, S. G. Shiah, C. J. Wang and S. E. Chuang, Suppression of apoptosis by Bcl-2 to enhance benzene metabolites-induced oxidative DNA damage and mutagenesis: A possible mechanism of carcinogenesis, *Mol. Pharmacol.*, 1999, **55**(5), 894.
61. L. Liang, M. S. Mendonca, L. Deng, S. C. Nguyen, C. Shao and J. A. Tischfield, Reduced apoptosis and increased deletion mutations at Aprt locus *in vivo* in mice exposed to repeated ionizing radiation, *Cancer Res.*, 2007, **67**(5), 1910.
62. A. L. Edinger and C. B. Thompson, Death by design: apoptosis, necrosis and autophagy, *Curr. Opin. Cell Biol.*, 2004, **16**(6), 663.
63. V. A. Fadok, D. L. Bratton and P. M. Henson, Phagocyte receptors for apoptotic cells: recognition, uptake, and consequences, *J. Clin. Invest.*, 2001, **108**(7), 957.
64. G. Petrovski, G. Zahuczky, K. Katona, G. Vereb, W. Martinet, Z. Nemes, W. Bursch and L. Fesus, Clearance of dying autophagic cells of different origin by professional and non-professional phagocytes, *Cell Death Differ.*, 2007, **14**(6), 1117.
65. P. Scaffidi, T. Misteli and M. E. Bianchi, Release of chromatin protein HMGB1 by necrotic cells triggers inflammation, *Nature*, 2002, **418**(6894), 191.
66. Y. Kondo and S. Kondo, Autophagy and cancer therapy, *Autophagy*, 2006, **2**(2), 85.
67. L. Yu, A. Alva, H. Su, P. Dutt, E. Freundt, S. Welsh, E. H. Baehrecke and M. J. Lenardo, Regulation of an ATG7-beclin 1 program of autophagic cell death by caspase-8, *Science*, 2004, **304**(5676), 1500.
68. S. Luo and D. C. Rubinsztein, Atg5 and Bcl-2 provide novel insights into the interplay between apoptosis and autophagy, *Cell Death Differ.*, 2007, **14**(7), 1247.
69. R. Zhang, J. Gong, H. Wang, L. Wang and L. W. Ran, Deoxycholate induces apoptosis in cultured normal human esophageal mucosal epithelial cells, *Di Yi Jun Yi Da Xue Xue Bao*, 2005, **25**(10), 1240.
70. M. J. Redlak, M. S. Dennis and T. A. Miller, Apoptosis is a major mechanism of deoxycholate-induced gastric mucosal cell death, *Am. J. Physiol. Gastrointest. Liver Physiol.*, 2003, **285**(5), G870.
71. D. N. Criddle, S. Gillies, H. K. Baumgartner-Wilson, M. Jaffar, E. C. Chinje, S. Passmore, M. Chvanov, S. Barrow, O. V. Gerasimenko, A. V. Tepikin, R. Sutton and O. H. Petersen, Menadione-induced reactive oxygen species generation via redox cycling promotes apoptosis of murine pancreatic acinar cells, *J. Biol. Chem.*, 2006, **281**(52), 40485.

72. T. Patel, S. F. Bronk and G. J. Gores, Increases of intra-cellular magnesium promote glycodeoxycholate-induced apoptosis in rat hepatocytes, *J. Clin. Invest.*, 1994, **94**(6), 2183.

73. P. Kwo, T. Patel, S. F. Bronk and G. J. Gores, Nuclear serine protease activity contributes to bile-acid-induced apoptosis in hepatocytes, *Am. J. Physiol.*, 1995, **268**, G613.

74. B. A. Jones, Y. P. Rao, R. T. Stravitz and G. J. Gores, Bile salt-induced apoptosis of hepatocytes involves activation of protein kinase C, *Am. J. Physiol.*, 1997, **272**, 1109.

75. T. Patel and G. J. Gores, Inhibition of bile-salt-induced hepatocyte apoptosis by the anti-oxidant lazaroid U83836E, *Toxicol. Appl. Pharmacol.*, 1997, **142**(1), 116.

76. W. A. Faubion, M. E. Guicciardi, H. Miyoshi, S. F. Bronk, P. J. Roberts, P. A. Svingen, S. H. Kaufmann and G. J. Gores, Toxic bile salts induce rodent hepatocyte apoptosis via direct activation of Fas, *J. Clin. Invest.*, 1999, **103**(1), 137.

77. T. Sodeman, S. F. Bronk, P. J. Roberts, H. Miyoshi and G. J. Gores, Bile salts mediate hepatocyte apoptosis by increasing cell surface trafficking of Fas, *Am. J. Physiol. Gastrointest. Liver Physiol.*, 2000, **278**(6), G992.

78. L. Qiao, E. Studer, K. Leach, R. McKinstry, S. Gupta, R. Decker, R. Kukreja, K. Valerie, P. Nagarkatti, W. El Deiry, J. Molkentin, R. Schmidt-Ullrich, P. B. Fisher, S. Grant, P. B. Hylemon and P. Dent, Deoxycholic acid (DCA) causes ligand-independent activation of epidermal growth factor receptor (EGFR) and FAS receptor in primary hepatocytes: inhibition of EGFR/mitogen-activated protein kinase-signalling module enhances DCA-induced apoptosis, *Mol. Biol. Cell.*, 2001, **12**(9), 2629.

79. B. Yerushalmi, R. Dahl, M. W. Devereaux, E. Gumpricht and R. J. Sokol, Bile-acid-induced rat hepatocyte apoptosis is inhibited by anti-oxidants and blockers of the mitochondrial permeability transition, *Hepatology*, 2001, **33**(3), 616.

80. H. Higuchi, S. F. Bronk, Y. Takikawa, N. Werneburg, R. Takimoto, W. El-Deiry and G. J. Gores, The bile acid glycochenodeoxycholate induces trail-receptor 2/DR5 expression and apoptosis, *J. Biol. Chem.*, 2001, **276**(42), 38610.

81. C. M. Rodrigues, G. Fan, X. Ma, B. T. Kren and C. J. Steer, A novel role for ursodeoxycholic acid in inhibiting apoptosis by modulating mitochondrial membrane perturbation, *J. Clin. Invest.*, 1998, **101**(12), 2790.

82. A. Hague, D. J. Elder, D. J. Hicks and C. Paraskeva, Apoptosis in colorectal tumour cells: induction by the short chain fatty acids butyrate, propionate and acetate and by the bile salt deoxycholate, *Int. J. Cancer*, 1995, **60**(3), 400.

83. M. C. Marchetti, G. Migliorati, R. Moraca, C. Riccardi, I. Nicoletti, R. Fabiani, V. Mastrandrea and G. Morozzi, Possible mechanisms involved in apoptosis of colon tumour cell lines induced by deoxycholic acid, short-chain fatty acids, and their mixtures, *Nutr. Cancer*, 1997, **28**(1), 74.

84. J. D. Martinez, E. D. Stratagoules, J. M. LaRue, A. A. Powell, P. R. Gause, M. T. Craven, C. M. Payne, M. B. Powell, E. W. Gerner and D. L. Earnest, Different bile acids exhibit distinct biological effects: the tumour promoter deoxycholic acid induces apoptosis and the chemopreventive agent ursodeoxycholic acid inhibits cell proliferation, *Nutr. Cancer*, 1998, **31**(2), 111.

85. D. Washo-Stultz, N. Hoglen, H. Bernstein, C. Bernstein and C. M. Payne, Role of nitric oxide and peroxynitrite in bile salt-induced apoptosis: relevance to colon carcinogenesis, *Nutr. Cancer*, 1999, **35**(2), 180.

86. D. Washo-Stultz, C. Crowley, C. M. Payne, C. Bernstein, S. Marek, E. W. Gerner and H. Bernstein, Increased susceptibility of cells to inducible apoptosis during growth from early to late log phase: an important caveat for *in vitro* apoptosis research, *Toxicol. Lett.*, 2000, **116**(3), 199.

87. A. I. Haza, B. Glinghammar, A. Grandien and J. Rafter, Effect of colonic luminal components on induction of apoptosis in human colonic cell lines, *Nutr. Cancer*, 2000, **36**(1), 79.

88. K. Schlottman, F. P. Wachs, R. C. Krieg, F. Kullmann, J. Scholmerich and G. Rogler, Characterization of bile salt-induced apoptosis in colon cancer cell lines, *Cancer Res.*, 2000, **60**(15), 4270.

89. S. Yui, T. Saeki, R. Kanamoto and K. Iwami, Characteristics of apoptosis in HCT116 colon cancer cells induced by deoxycholic acid, *J. Biochem. (Tokyo)*, 2005, **138**(2), 151.

90. F. P. Wachs, R. C. Krieg, C. M. Rodrigues, H. Messmann, F. Kullmann, R. Knuchel-Clarke, J. Scholmerich, G. Rogler and K. Schlottmann, Bile salt-induced apoptosis in human colon cancer cell lines involves the mitochondrial transmembrane potential but not the CD95 (Fas/Apo-1) receptor, *Int. J. Colorectal Dis.*, 2005, **20**(2), 103.

91. H. S. Samaha, C. Bernstein, C. M. Payne, H. Garewal, R. Sampliner and H. Bernstein, Bile salt induction of apoptosis in goblet cells of the normal human colonic mucosa: relevance to colon cancer, *Acta Microsc.*, 1995, **4**, 43.

92. C. L. Crowley, C. M. Payne, H. Bernstein, C. Bernstein and D. Roe, The NAD+ precursors, nicotinic acid and nicotinamide protect cells against apoptosis induced by a multiple stress inducer, deoxycholate, *Cell Death Differ.*, 2000, **7**(3), 314.

93. D. Nehra, P. Howell, C. P. Williams, J. K. Pye and J. Beynon, Toxic bile acids in gastro-oesophageal reflux disease: influence of gastric acidity, *Gut*, 1999, **44**(5), 598.

94. M. Menges, M. Muller and M. Zeitz, Increased acid and bile reflux in Barrett's esophagus compared to reflux esophagitis, and effect of proton pump inhibitor therapy, *Am. J. Gastroenterol.*, 2001, **96**(2), 331.

95. H. J. Stein, W. K. Kauer, H. Feussner and J. R. Siewert, Bile reflux in benign and malignant Barrett's esophagus: effect of medical acid suppression and nissen fundoplication, *J. Gastrointest. Surg.*, 1998, **2**(4), 333.

96. Y. Su, X. Chen, M. Klein, M. Fang, S. Wang, C. S. Yang and R. K. Goyal, Phenotype of columnar-lined esophagus in rats with

esophagogastroduodenal anastomosis: similarity to human Barrett's esophagus, *Lab. Invest.*, 2004, **84**(6), 753.
97. M. Fein, J. H. Peters, P. Chandrasoma, A. P. Ireland, S. Oberg, M. P. Ritter, C. G. Bremner, J. A. Hagen and T. R. DeMeester, Duodenoeso-phageal reflux induces esophageal adenocarcinoma without exogenous carcinogen, *J. Gastrointest. Surg.*, 1998, **2**(3), 260.
98. A. Kuwahara, T. Saito and M. Kobayashi, Bile acids promote carcino-genesis in the remnant stomach of rats, *J. Cancer Res. Clin. Oncol.*, 1989, **115**(5), 423.
99. K. Kondo, Duodenogastric reflux and gastric stump carcinoma, *Gastric Cancer*, 2002, **5**(1), 16.
100. O. Kobori, T. Shimizu, M. Maeda, Y. Atomi, J. Watanabe, M. Shoji and Y. Morioka, Enhancing effect of bile and bile acid on stomach tumouri-genesis induced by N-methyl-N'-nitro-N-nitrosoguanidine in Wistar rats, *J. Natl. Cancer Inst.*, 1984, **73**(4), 853.
101. O. N. Tucker, A. J. Dannenberg, E. K. Yang and T. J. Fahey III, Bile acids induce cyclo-oxygenase-2 expression in human pancreatic cancer cell lines, *Carcinogenesis*, 2004, **25**(3), 419.
102. T. Adachi, Y. Tajima, T. Kuroki, T. Mishima, A. Kitasato, K. Fukuda, R. Tsutsumi and T. Kanematsu, Bile-reflux into the pancreatic ducts is associated with the development of intra-ductal papillary carcinoma in hamsters, *J. Surg. Res.*, 2006, **136**(1), 106.
103. A. S. Knisely, S. S. Strautnieks, Y. Meier, B. Stieger, J. A. Byrne, B. C. Portmann, L. N. Bull, L. Pawlikowska, B. Bilezikci, F. Ozcay, A. Laszlo, L. Tiszlavicz, L. Moore, J. Raftos, H. Arnell, B. Fischler, A. Nemeth, N. Papadogiannakis, J. Cielecka-Kuszyk, I. Jankowska, J. Pawlowska, H. Melin-Aldana, K. M. Emerick, P. F. Whitington, G. Mieli-Vergani and R. J. Thompson, Hepatocellular carcinoma in ten children under five years of age with bile-salt export pump deficiency, *Hepatology*, 2006, **44**(2), 478.
104. R. K. Ross, N. M. Hartnett, L. Bernstein and B. E. Henderson, Epi-demiology of adenocarcinomas of the small intestine: is bile a small bowel carcinogen? *Br. J. Cancer*, 1991, **63**(1), 143.
105. D. G. Jagelman, J. J. DeCosse and H. J. Bussey, Upper gastro-intestinal cancer in familial adenomatous polyposis, *Lancet*, 1988, **1**(8595), 1149.
106. A. D. Spigelman, C. B. Williams, I. C. Talbot, P. Domizio and R. K. Phillips, Upper gastro-intestinal cancer in patients with familial adeno-matous polyposis, *Lancet*, 1989, **2**(8666), 783.
107. N. N. Mahmoud, A. J. Dannenberg, R. T. Bilinski, J. R. Mestre, A. Chadburn, M. Churchill, C. Martucci and M. M. Bertagnolli, Ad-ministration of an unconjugated bile acid increases duodenal tumours in a murine model of familial adenomatous polyposis, *Carcinogenesis*, 1999, **20**(2), 299.
108. M. J. Hill, Bile flow and colon cancer, *Mutat. Res.*, 1990, **238**(3), 313.
109. M. J. Hill, B. S. Drasar, G. Hawksworth, V. Aries, J. S. Crowther and R. E. Williams, Bacteria and aetiology of cancer of large bowel, *Lancet*, 1971, **1**(769), 95.

110. M. J. Hill, A. J. Taylor, M. H. Thompson and R. Wait, Fecal steroids and urinary volatile phenols in four Scandinavian populations, *Nutr. Cancer*, 1982, **4**(1), 67.

111. J. S. Crowther, B. S. Drasar, M. J. Hill, R. Maclennan, D. Magnin, S. Peach and C. H. Teoh-chan, Faecal steroids and bacteria and large bowel cancer in Hong Kong by socio-economic groups, *Br. J. Cancer*, 1976, **34**(2), 191.

112. B. S. Reddy and E. L. Wynder, Large-bowel carcinogenesis: fecal constituents of populations with diverse incidence rates of colon cancer, *J. Natl. Cancer Inst.*, 1973, **50**(6), 1437.

113. B. S. Reddy and E. L. Wynder, Metabolic epidemiology of colon cancer. Fecal bile acids and neutral sterols in colon cancer patients and patients with adenomatous polyps, *Cancer*, 1977, **39**(6), 2533.

114. B. S. Reddy, A. R. Hedges, K. Laakso and E. L. Wynder, Metabolic epidemiology of large bowel cancer: fecal bulk and constituents of high-risk North American and low-risk Finnish population, *Cancer*, 1978, **42**(6), 2832.

115. O. M. Jensen, R. MacLennan and J. Wahrendorf, Diet, bowel function, fecal characteristics, and large bowel cancer in Denmark and Finland, *Nutr. Cancer*, 1982, **4**(1), 5.

116. L. Domellof, L. Darby, D. Hanson, L. Mathews, B. Simi and B. S. Reddy, Fecal sterols and bacterial beta-glucuronidase activity: a preliminary metabolic epidemiology study of healthy volunteers from Umea, Sweden, and metropolitan New York, *Nutr. Cancer*, 1982, **4**(2), 120.

117. T. Narisawa, N. E. Magadia, J. H. Weisburger and E. L. Wynder, Promoting effect of bile acids on colon carcinogenesis after intra-rectal instillation of N-methyl-N'-nitro-N-nitrosoguanidine in rats, *J. Natl. Cancer Inst.*, 1974, **53**(4), 1093.

118. B. S. Reddy, T. Narasawa, J. H. Weisburger and E. L. Wynder, Promoting effect of sodium deoxycholate on colon adenocarcinomas in germfree rats, *J. Natl. Cancer Inst.*, 1976, **56**(2), 441.

119. B. S. Reddy, K. Watanabe, J. H. Weisburger and E. L. Wynder, Promoting effect of bile acids in colon carcinogenesis in germ-free and conventional F344 rats, *Cancer Res.*, 1977, **37**(9), 3238.

120. C. K. McSherry, B. I. Cohen, V. D. Bokkenheuser, E. H. Mosbach, J. Winter, N. Matoba and J. Scholes, Effects of calcium and bile-acid feeding on colon tumours in the rat, *Cancer Res.*, 1989, **49**(21), 6039.

121. S. Lechner, U. Muller-Ladner, K. Schlottmann, B. Jung, M. McClelland, J. Ruschoff, J. Welsh, J. Scholmerich and F. Kullmann, Bile acids mimic oxidative stress induced up-regulation of thioredoxin reductase in colon cancer cell lines, *Carcinogenesis*, 2002, **23**(8), 1281.

122. R. J. Sokol, B. M. Winklhofer-Roob, M. W. Devereaux and J. M. McKim, Jr., Generation of hydroperoxides in isolated rat hepatocytes and hepatic mitochondria exposed to hydrophobic bile acids, *Gastroenterology*, 1995, **109**(4), 1249.

123. R. J. Sokol, M. S. Straka, R. Dahl, M. W. Devereaux, B. Yerushalmi, E. Gumpricht, N. Elkins and G. Everson, Role of oxidant stress in the permeability transition induced in rat hepatic mitochondria by hydrophobic bile acids, *Pediatr. Res.*, 2001, **49**(4), 519.

124. R. Reinehr, S. Becker, V. Keitel, A. Eberle, S. Grether-Beck and D. Haussinger, Bile salt-induced apoptosis involves NADPH oxidase isoform activation, *Gastroenterology*, 2005, **129**(6), 2009.

125. P. A. Craven, J. Pfanstiel and F. R. DeRubertis, Role of reactive oxygen in bile salt stimulation of colonic epithelial proliferation, *J. Clin. Invest.*, 1986, **77**(3), 850.

126. N. Katada, R. A. Hinder, T. C. Smyrk, N. Hirabayashi, G. Perdikis, R. J. Lund, T. Woodward and P. J. Klingler, Apoptosis is inhibited early in the dysplasia-carcinoma sequence of Barrett esophagus, *Arch. Surg.*, 1997, **132**(7), 728.

127. S. Badvie, A. Hanna-Morris, H. J. Andreyev, P. Cohen, S. Saini and T. G. Allen-Mersh, A "field change" of inhibited apoptosis occurs in colorectal mucosa adjacent to colorectal adenocarcinoma, *J. Clin. Pathol.*, 2006, **59**(9), 942.

CHAPTER 4
The Genotoxicity of Bile Acids

LAURA J. HARDIE

Molecular Epidemiology Unit, Leeds Institute for Genetics, Health and Therapeutics, LIGHT Laboratories, Clarendon Way, University of Leeds, Leeds LS2 9JT, UK

4.1   Introduction

Maintaining the integrity of DNA structure and function is critical to normal cellular function. Cells are continuously exposed to DNA damage from endogenous (*e.g.* free radicals) and exogenous sources (*e.g.* environmental toxins, ionising radiation). In most instances damaged DNA is repaired *via* a complex repair strategy.[1] However, this system is not error-free and unrepaired or misrepaired DNA damage may give rise to fixed genetic changes such as mutations, deletions, amplifications or chromosomal rearrangements. If these changes impair the function of key tumour-suppressor genes, cause oncogene activation or otherwise disrupt the regulation of key pathways linked to cell proliferation, differentiation, growth or apoptosis, this increases the likelihood of malignant transformation.

Many sites of exposure to bile in the body are associated with the development of cancer, *e.g.* the oesophagus, gallbladder and bile duct, pancreas, small intestine and colon (reviewed in ref. 2). One explanation for increased cancer at these sites could be that bile acids stimulate carcinogenesis *via* DNA-damaging effects. This chapter provides an overview of research conducted in relation to establishing the genotoxic and carcinogenic effects of bile acids.

4.2   Definitions

A genotoxin can be defined as any substance that causes damage to cellular DNA. This may occur *via* alterations to the DNA structure, sequence or

Issues in Toxicology
Bile Acids: Toxicology and Bioactivity
Edited by Gareth Jenkins and Laura J. Hardie
© Royal Society of Chemistry, 2008

segregation of genetic material. Genotoxic agents that give rise to permanent and therefore potentially transmissible changes in the DNA sequence (mutations) are specifically referred to as mutagens or mutagenic.

In animal studies, most potent mutagens are also found to be carcinogenic. The International Agency for Research on Cancer (IARC) defines compounds as "carcinogenic" if they are able to increase the incidence of malignant tumours, reduce their time to development, or increase their severity or multiplicity (http://monographs.iarc.fr/ENG/Preamble/index.php). Carcinogens may have genotoxic or nongenotoxic modes of action. Genotoxic carcinogens stimulate carcinogenesis by direct interaction with DNA, resulting in DNA damage or chromosomal aberrations, whereas nongenotoxic carcinogens do not directly interact with DNA but alter DNA function *via* epigenetic mechanisms such as altering gene expression or signal transduction.[3]

4.3   Bile Acids as Carcinogens

The interest in bile acids as potential carcinogens was subject to investigation as early as 1940 when Cook *et al.* reported in Nature that repeated injection of deoxycholic acid into the flanks of mice could induce tumour formation in mice.[4] Furthermore, Kelsey and Pienta showed that treatment of hamster embryo cells with lithocholic acid could cause cell transformation.[5]

However, subsequent studies did not find clear evidence to support the view that bile acids could independently stimulate tumour formation utilising rat models. Rather, the findings indicated that bile acids could enhance the effect of other carcinogens in these models.[6-10] An example of such a study is by McSherry *et al.* in which male Fischer rats were fed diets supplemented with cholic acid (0.2%) and administered the colonic carcinogen, *N*-Methyl-*N*-nitrosurea (MNU), intra-rectally.[6] Fifty-five per cent of MNU treated rats on standard diet developed tumours, a figure that increased to eighty per cent in MNU-treated rats given dietary cholic acid. Rats fed cholic acid supplemented diet alone did not develop tumours.

Such results led many researchers to suggest that bile acids act primarily as promoting agents once a tumour is initiated; probably *via* effects on modulating cell survival or proliferation. However, other studies suggest that this interpretation may be too simplistic. Mahmoud and coworkers examined the effect of the unconjugated bile acid, chenodeoxycholate in the Min/+ mouse model.[11] These mice have a germline mutation in the *Apc* gene and develop multiple tumours predominantly in the small intestine, providing a model for the condition, familial adenomatous polyposis. The authors demonstrated that dietary supplementation with chenodeoxycholate caused a nine-fold increase in the number of duodenal tumours that developed in these mice; an effect that occurred in the absence of administration of a classic exogenous carcinogen such as MNU.

Similarly, Kitazawa and colleagues examined the effect of deoxycholic acid (DCA) and lithocholic acid (LCA) during the initiation of rat hepatocarcinogenesis.[12] Administration of DCA in the diet, in advance of, and for the week following partial hepatectomy, resulted in increased numbers of hyper-plastic liver nodules.

More recently, Guy *et al.* fed C57BL/6 mice a zinc-deficient diet containing deoxycholic acid for up to 152 days. Sixty-three per cent of mice on this diet developed oesophageal lesions that resembled the preneoplastic precursor for oesophageal adenocarcinoma, Barrett's oesophagus.[13]

These latter studies highlight the possibility that bile acids may have carcinogenic or co-carcinogenic effects but these may only be manifest at particular sites under certain physiological conditions. In addition, Bernstein *et al.*,[2] have argued that the apparent lack of carcinogenic effects of bile acids in some murine studies might also be explained by i) different gut and bile acid physiology in these animals compared with man, ii) inadequate exposure times and/or that iii) the lifespan of rodents is not sufficiently long to observe effects; particularly if bile acids have weak carcinogenic effects.

If bile acids do have carcinogenic effects *in vivo*, one would expect these compounds to display certain genotoxic and carcinogenic effects utilising established *in-vitro* assay systems.

4.4   DNA Adducts

Carcinogens often interact and bind directly with DNA. For this reason, investigators sought to establish whether bile acids could directly induce DNA adducts. Utilising the ^{32}P-post labelling technique, Scates *et al.*[14,15] reported that human bile was capable of inducing DNA adducts in cell lines and *in vivo* in rats. Higher levels of adduct formation were reported utilising bile collected from patients with familial adenomatous polyposis compared with control subjects. Several studies subsequently assessed whether the DNA adducts formed from bile could be attributed to specific bile acids. Hamada *et al.*[16] reported that unconjugated chenodeoxycholic and lithocholic acids were able to bind to naked DNA *in vitro*, although notably, they could not induce DNA adduct formation by these compounds either *in vitro* in the Hep G2 cell line or in Fischer rats. Scates and colleagues subsequently challenged Hamada's report that bile acids could form adducts on naked DNA, suggesting that the adduct formation observed in their study could have arisen as an artefact of the ^{32}P post-labelling technique.[17,18] These studies have therefore provided little evidence to support the formation of bile-acid adducts to DNA. Nevertheless, there is increasing evidence to support damaging effects of bile acids on DNA and this is now discussed.

4.5   Bile Acids cause DNA Damage

The SOS response is an inducible DNA-repair system that allows bacteria to survive rapid increases in DNA damage.[19] Two proteins play a key role in regulating this response: a repressor, LexA and the inducer, RecA filament. In the uninduced state, the LexA repressor is bound to the SOS box present in the promoter region of SOS genes and inhibits their expression. Following DNA damage, the RecA protein binds to single-stranded DNA, forming a nucleoprotein filament. This filament can initiate homologous recombination by invading a homologous double-stranded DNA sequence and catalysing strand exchange. The filament also stimulates cleavage of the LexA repressor, and therefore de-represses SOS genes. Bernstein *et al.*,[20] showed that in *E. coli*, deoxycholate, chenodeoxycholate, ursodeoxycholate and glycocholate induce the oxidative SOS responsive genes; micF and osmY, and stimulate expression of DinD a protein that is specifically induced by DNA damage. Prieto *et al.*[21] also reported RecA-dependent induction of the SOS response in *Salmonella enterica*, following exposure to a range of bile salts.

Kandell and Bernstein[22] published one of the earliest reports to suggest that bile acids also demonstrate DNA-damaging effects in eukaryotic cells. They showed that human foreskin fibroblasts underwent unscheduled DNA synthesis (indicating DNA repair), as measured by tritiated thymidine incorporation when cells were treated with increasing concentrations of sodium deoxycholate or chenodeoxycholate. Utilising mutant Chinese hamster ovary cells deficient in strand rejoining (EM9), the authors were able to demonstrate that the repair of deoxycholate-induced DNA damage was dependent on strand break repair capacity.

PARP binds to DNA surrounding single- or double-strand breaks and attaches polymers of ADP-ribose to itself and other proteins. This increases the negative charge in this area, thus facilitating DNA repair. Payne *et al.*,[23] reported increased poly(ADP-ribose) polymerase (PARP) activity in Jurkat cells treated with sodium deoxycholate. Activation of PARP in colonic cells was similarly reported by Glinghammer *et al.*[24] following treatment with this bile acid.

Additional evidence to directly support the formation of strand breaks by bile acids, came from the application of the single-cell gel electrophoresis or comet assay (for a full review of this technique see[25]). This assay allows the measurement of DNA damage in individual cells. The underlying principle of the assay is that under appropriate conditions DNA strand breaks cause relaxation of supercoiled DNA in the nucleus and relaxed strands are able to migrate towards the anode during electrophoresis. In cells with DNA damage, this can be visualised as a "comet" shape, with undamaged DNA remaining in the "head" of the comet and damaged DNA migrating to form the comet "tail". The amount of DNA that migrates to the comet tail is directly proportional to the level of strand-break damage. A schematic of the technique is depicted in Figure 4.1. If the assay is conducted under neutral lysis and electrophoresis conditions, double-strand breaks are detected; whereas under alkali lysis conditions, a combination of DNA double-strand breaks, single-strand breaks and alkali labile sites (*e.g.* abasic sites) can be measured.

Utilising the alkaline comet assay Jolly *et al.*[26] showed that exposing the nontumourigenic squamous oesophageal cell line, Het-1A to deoxycholate at

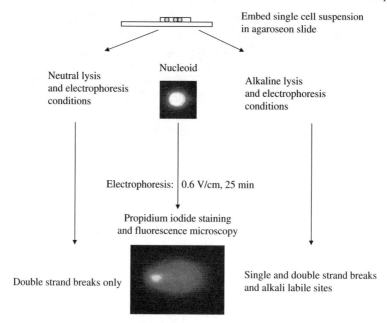

Embed single cell suspension
in agaroseon slide

Nucleoid

Neutral lysis
and electrophoresis
conditions

Alkaline lysis
and electrophoresis
conditions

Electrophoresis: | 0.6 V/cm, 25 min

Propidium iodide staining
and fluorescence microscopy

Double strand breaks only

Single and double strand breaks
and alkali labile sites

Figure 4.1   The comet assay. A single-cell suspension is embedded in agarose on a slide. Cells are then subject to lysis followed by electrophoresis. If present, damaged DNA migrates out of the nucleoid structure during electrophoresis to producing a characteristic comet shape. Double-strand breaks are revealed under neutral conditions, whereas alkali conditions additionally show single-strand breaks and alkali labile sites. Image analysis of stained DNA is used to quantitate the amount of damaged DNA in the comet tail.

concentrations that are physiologically relevant during episodes of reflux caused DNA damage in a dose-dependent manner at neutral pH. Interestingly, strand-break damage could not be detected with deoxycholate treatment under acidic pH conditions (pH 4.5). The detection of DNA strand breaks has also been reported in studies of colonic epithelial cell lines treated with deoxycholate using this technique.[27,28]

4.6   The Role of Reactive Oxygen and Nitrogen Species

The question arises as to how bile acids stimulate DNA damage if they do not adduct to DNA (discussed above) or cause direct strand-break damage to naked DNA.[27] One possibility is that cellular products generated as a consequence of bile-acid exposure effect the DNA-damaging action of bile acids. In this regard, bile acids have been shown to damage both plasma membranes and mitochondria, stimulating the release of reactive oxygen species.[29]

Craven and coworkers[30,31] demonstrated increased free-radical production, as measured by chemiluminescence, in colonic crypts from rats treated with deoxycholic acid and glycochenodeoxycholic acid has been shown to stimulate reactive oxygen species (ROS) production in mouse cholangiocytes.[31] Jenkins *et al.*[32] reported increased ROS release following deoxycholate treatment in oesophageal adenocarcinoma cells as measured utilising fluorescent ROS-sensitive probes and increased levels of carbonyl residues on proteins.

Most recently, Dvorak and colleagues[33] demonstrated that treatment of oesophageal cells with bile-acid cocktail at low pH (4.0) stimulates superoxide anion release from mitochondria. This was associated with increased formation of the oxidative DNA lesion, 8-hydroxydeoxyguanosine in oesophageal cell lines and Barrett's biopsies; effects that were not observed with bile acids at higher pH.

In addition to reactive oxygen species, reactive nitrogen species may also play a key role in bile-stimulated DNA damage. Jenkins *et al.*[34] have shown that bile acids activate NF-κB at neutral pH. NF-κB increases the expression of a number of downstream target genes including inducible nitric oxide synthase (iNOS). Increased levels of nitrosylated proteins have been reported following incubation of the colon cell line (HCT-116RC) with deoxycholate.[35] We have recently observed in our laboratory that treatment of the oesophageal cell line Het-1A, with deoxycholic acid at neutral pH stimulates the formation of nitric oxide (NO) radicals as measured by increased nitrite accumulation in culture medium and increased intra-cellular fluorescence of the NO specific probe, DAF-FM diacetate. Moreover, incubation with a NO scavenger (C-PTIO), inhibitor of inducible NO synthase (L-NAME) or NF-κB inhibitor (TPCK), all protected against DNA damage stimulated by deoxycholic acid at neutral pH, as measured by strand breaks in the comet assay, (Dr Jim Jolly, personal communication). These data provide direct evidence that NO radicals are generated in oesophageal cells following deoxycholic acid treatment *via* NF-κB activation, and play an important role in mediating strand-break damage at neutral pH.

Together these data suggest that a range of free-radical species may be generated in response to deoxycholic acid exposure in tissues. The particular radical species formed may be dependent upon specific cellular conditions *e.g.* pH, antioxidant status or metal ion levels. Acidic pH may promote the formation of 8-hydroxydeoxyguanosine residues *via* superoxide anion production, whereas neutral conditions may favour DNA strand-break formation mediated by nitric oxide. The nature and distribution of DNA damage arising under different conditions of bile exposure may have important implications for the consequences of bile exposure and is an area that requires more detailed investigation.

4.7 Mutagenic Effects of Bile Acids

Given that bile acids, particularly deoxycholic acid, can damage DNA, one might expect bile acids to demonstrate mutagenic effects. This might be particularly apparent in eukaryotic cells that could sustain mitochondrial damage or release NO *via* the mechanisms described above.

Probably the most widely recognised and validated test to assess the mutagenic capacity of compounds *in vitro* is the *Salmonella typhimurium*/microsome assay or Ames test.[36] In this test, strains of *Salmonella* have been selected that contain mutations in the histidine operon (base-pair substitution or frameshift mutants) leaving the bacteria unable to synthesise histidine. The presence of histidine in culture media is therefore a requirement for growth of these strains. Exposing these strains to mutagens can restore histidine-independent growth if the mutations generated cause a reversion event in the appropriate DNA sequence. These reversion mutants can subsequently be detected by their ability to grow in histidine-depleted media.

The assay is usually conducted in a plate form, with revertant colonies stained and counted. However, to increase the assay's sensitivity for weaker mutagens, the fluctuation assay can be conducted. In this modification, bacteria treated with the suspect mutagen are grown in suspension culture rather than plated out. Increased turbidity of the culture medium over time is used to detect the presence of revertants.

Utilising the standard Ames test, studies have reported a lack of mutagenic activity with bile acids or their derivatives.[37,38] However, by applying the more sensitive fluctuation assay, Watabe and Bernstein[39] reported mutagenic effects with cholic, chenodeoxycholic, deoxycholic and ursodeoxycholic but not lithocholic acid using *Salmonella typhinurium* tester strains TA100 and TA98. These mutant strains detect base-pair substitutions and frameshift mutations, respectively. Wilpart and Roberfroid[40] subsequently reported that lithocholic acid and deoxycholic acid could also enhance the frequency of revertants in *Salmonella typhinurium* by several mutagens (dimethylhydrazine, 2-nitrofluorene, *N*-methyl-*N'*-nitro-*N*-nitrosoguanidine and 2-acetyl-aminofluorene) suggesting co-mutagenic effects.

Utilising a reversion assay in *Salmonella enterica*, Prieto *et al.*[21,41] reported an increased frequency of point mutations following bile-salt exposure. Mutations were predominantly nucleotide substitutions (GC to AT transitions) and -1 frameshift mutations.[21,41] The frameshifts were dependent on SOS induction and linked to the activity of DinB polymerase (Pol IV). The authors proposed that the GC to AT transitions stimulated by bile, could have arisen from oxidative processes giving rise to oxidised cytosine residues. Consistent with this hypothesis, the authors demonstrated that strains of *S. enterica*-lacking enzymes required for base-excision repair (endonuclease III and exonuclease IV) and the removal of oxidised bases, demonstrated increased bile-acid sensitivity compared with competent strains. In another study using *E. coli*, resistance to the DNA-damaging effects of bile was associated with Dam-directed mismatch repair, a pathway also involved with the repair of oxidative DNA lesions.[42]

Jenkins *et al.*[43] examined the formation of mutations in the p53 gene in oesophageal cell lines following treatment with deoxycholic acid for 24 h. The authors demonstrated an increased frequency of GC to AT mutations in KYSE-30 cells at neutral pH. The IARC TP53 Mutation Database compiles (http://www-p53.iarc.fr/index.html) p53 gene mutations identified in human

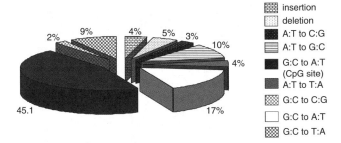

Figure 4.2 The spectrum of p53 mutations detected in oesophageal adenocarcinoma tumours. Data were downloaded from the IARC database and based on an analysis of 260 separate mutation profiles.

cancers and cell lines that have been reported in the peer-reviewed literature.[44] Analysis of p53 mutations from colon cancers and oesophageal adenocarcinomas in the IARC p53 database reveals that GC to AT mutations are present at a high frequency in this tumour type which may be consistent with bile exposure (Figure 4.2).

Together these data suggest that bile acids may induce mutations in DNA, and reinforces the notion that bile acids may play a key role as carcinogens or co-carcinogens in the gastro-intestinal tract.

4.8 Bile Acids and Chromosomal Damage

Micronuclei are DNA-containing bodies that are expressed in dividing cells that contain chromosome breaks lacking centromeres and/or whole chromosomes that are unable to migrate to spindle poles during mitosis (reviewed in[45]). Micronuclei are visualised and scored in mitotic binucleate cells that have been blocked from undergoing cytokinesis with cytochalasin B. Kinetochore staining (staining of centromeres) can be used to distinguish micronuclei that contain whole chromosomes from chromosome fragments. This staining can therefore differentiate between clastogenic (chromosomal fragmentation) and aneugenic, genotoxic effects.

Jenkins *et al.*[32] demonstrated that the secondary bile acid, deoxycholic acid could induce micronuclei formation in the oesophageal adenocarcinoma cell line, OE33. The induction of micronuclei demonstrated a dose-dependent effect and occurred under both neutral and acidic pH conditions. An example of a micronucleus induced by treatment of the OE33 oesophageal adenocarcinoma cell line with deoxycholic acid is shown in Figure 4.3.

The majority of micronuclei formed in this study were kinetochore negative, indicating a predominantly clastogenic effect of deoxycholic acid in these cells. Clastogenicity has previously been reported for organic fractions from human faeces and for ursodeoxycholic acid, utilising Chinese hamster ovary cells and human lymphocytes.[46,47]

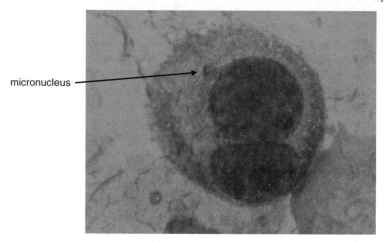

Figure 4.3 Example of a micronucleus formed following deoxycholic acid treatment of the OE33 oesophageal adenocarcinoma cell line (OE33). Courtesy of Dr Gareth Jenkins, University of Swansea.

4.9 Chemoprevention Strategies

There is accumulating evidence that the bile acid, deoxycholic acid, causes DNA damage and has mutagenic and clastogenic effects. These effects appear to be mediated in part, *via* the formation of free-radical species. Jenkins *et al.*[43] recently showed that free-radical production in oesophageal cells following deoxycholic acid treatment could be inhibited by coincubating cells with vitamin C *in vitro*. Pre-incubation or co-incubation with vitamin C *in vitro* also inhibited the formation of micronuclei formation in deoxycholic-acid-treated oesophageal cells. These results suggest it may be possible to protect against the genotoxic effects of bile acids by augmenting anti-oxidant levels in certain high-risk patient groups. This idea is supported by epidemiological studies that suggest that high levels of fruit and vegetable consumption, and diets particularly rich in vitamin C, protect against oesophageal adenocarcinoma development.[48–50] Given these links, and the elevated levels of DNA damage measured in Barrett's oesophagus tissue,[51] this has prompted vitamin C and curcumin intervention studies in Barrett's oesophagus patients that are now nearing completion.

4.10 Summary and Concluding Remarks

Bile acids have previously been considered to have primarily promoting effects during tumourigenesis. However, there is increasing evidence that certain bile acids, particularly deoxycholic acid, have significant genotoxic effects that could contribute to tumour initiation. Deoxycholic acid demonstrates genotoxic effects at both acidic pH and neutral conditions. In the context of reflux disease, this could explain why, despite widespread usage of acid-suppressing medication, the incidence of oesophageal adenocarcinoma continues to rise. The genotoxic effects of deoxycholic acid appear to be mediated *via* the release of free-radical species as a consequence of cell signalling and cellular damage stimulated by bile acids.

Limiting the genotoxicity of bile acids might be afforded by interventions including anti-oxidant supplementation in susceptible patient groups.

References

1. W. K. Hansen and M. R. Kelley, *J. Pharmacol. Exp. Ther.*, 2000, **295**, 1.
2. H. Bernstein, C. Bernstein, C. M. Payne, K. Dvorakova and H. Garewal, *Mutat Res. -Reviews in Mutat. Res.*, 2005, **589**, 47.
3. J. Ashby and I. F. H. Purchase, *Environ. Health Perspect*, 1992, **98**, 223.
4. J. W. Cook, E. I. Kennaway and N. M. Kennaway, *Nature*, 1940, **145**, 627.
5. M. I. Kelsey and R. J. Pienta, *Cancer Lett.*, 1979, **6**, 143.
6. C. K. McSherry, B. I. Cohen, V. D. Bokkenheuser, E. H. Mosbach, J. Winter, N. Matoba and J. Scholes, *Cancer Res.*, 1989, **49**, 6039.
7. B. S. Reddy, K. Watanabe, J. H. Weisburger and E. L. Wynder, *Cancer Res.*, 1977, **37**, 3238.
8. R. G. Cameron, K. Imaida, H. Tsuda and N. Ito, *Cancer Res.*, 1982, **42**, 2426.
9. H. Tsuda, T. Masui, K. Imaida, S. Fukushima and N. Ito, *Gann*, 1984, **75**, 871.
10. T. Narisawa, N. E. Magadia, J. H. Weisburg and E. L. Wynder, *J. Natl. Cancer Inst.*, 1974, **53**, 1093.
11. N. N. Mahmoud, A. J. Dannenberg, R. T. Bilinski, J. R. Mestre, A. Chadburn, M. Churchill, C. Martucci and M. M. Bertagnolli, *Carcinogenesis*, 1999, **20**, 299.
12. S. Kitazawa, A. Denda, M. Tsutsumi, T. Tsujiuchi, K. Hasegawa, K. Tamura, H. Maruyama and Y. Konishi, *Carcinogenesis*, 1990, **11**, 1323.
13. N. C. Guy, H. Garewal, H. Holubec, H. Bernstein, C. M. Payne, C. Bernstein, A. K. Bhattacharyya and K. Dvorak, *Nutr. Cancer*, 2007, **59**, 217.
14. D. K. Scates, A. D. Spigelman, R. K. S. Phillips and S. Venitt, *Carcinogenesis*, 1992, **13**, 731.
15. D. K. Scates, A. D. Spigelman, R. K. S. Phillips and S. Venitt, *Carcinogenesis*, 1996, **17**, 707.
16. K. Hamada, A. Umemoto, A. Kajikawa, M. J. Seraj and Y. Monden, *Carcinogenesis*, 1994, **15**, 1911.
17. D. K. Scates, A. D. Spigelman and S. Venitt, *Carcinogenesis*, 1994, **15**, 2945.
18. D. K. Scates, A. D. Spigelman and S. Venitt, *Carcinogenesis*, 1995, **16**, 1489.
19. B. Michel, *Plos. Biol.*, 2005, **3**, 1174.

20. C. Bernstein, H. Bernstein, C. M. Payne, S. E. Beard and J. Schneider, *Curr. Microbiol.*, 1999, **39**, 68.
21. A. I. Prieto, F. Ramos-Morales and J. Casadesus, *Genetics*, 2004, **168**, 1787.
22. R. L. Kandell and C. Bernstein, *Nutr. Cancer*, 1991, **16**, 227.
23. C. M. Payne, C. Crowley, D. Washo-Stultz, M. Briehl, H. Bernstein, C. Bernstein, S. Beard, H. Holubec and J. Warneke, *Cell Death Differ.*, 1998, **5**, 623.
24. B. Glinghammar, H. Inoue and J. J. Rafter, *Carcinogenesis*, 2002, **23**, 839.
25. P. L. Olive and J. P. Banath, *Nat. Proto.*, 2006, **1**, 23.
26. A. J. Jolly, C. P. Wild and L. J. Hardie, *Mutagenesis*, 2004, **19**, 319.
27. B. Glinghammar, H. Inoue and J. J. Rafter, *Carcinogenesis*, 2002, **23**, 839.
28. A. Powolny, J. Xu and G. Loo, *Int. J. Biochem. Cell. Biol.*, 2001, **33**, 193.
29. T. J. Sokol, R. Dahl, M. W. Devereaux, B. Yerushalmi, G. E. Kobak and E. Gumpricht, *J. Pediatr Gastroenterol Nutr.*, 2005, **41**, 235.
30. P. A. Craven, J. Pfanstiel and F. R. Derubertis, *J. Clin. Invest.*, 1986, **77**, 850.
31. D. Komichi, S. Tazuma, T. Nishioka, H. Hyogo and K. Chayama, *Free Radic. Biol. Med.*, 2005, **39**, 1418.
32. G. J. S. Jenkins, F. R. D'Souza, S. H. Suzen, Z. S. Eltahir, S. A. James, J. M. Parry, P. A. Griffiths and J. N. Baxter, *Carcinogenesis*, 2007, **28**, 136.
33. K. Dvorak, C. M. Payne, M. Chavarria, L. Ramsey, B. Dvorakova, H. Bernstein, H. Holubec, R. E. Sampliner, N. Guy, A. Condon, C. Bernstein, S. B. Green, A. Prasad and H. S. Garewal, *Gut*, 2007, **56**, 763.
34. G. J. S. Jenkins, K. Harries, S. H. Doak, A. Wilmes, A. P. Griffiths, J. N. Baxter and J. M. Parry, *Carcinogenesis*, 2004, **25**, 317.
35. M. Dall'Agnol, C. Bernstein, H. Bernstein, H. Garewal and C. M. Payne, *Proteomics*, 2006, **6**, 1654.
36. B. N. Ames, J. Mccann and E. Yamasaki, *Mutat Res.*, 1975, **31**, 347.
37. S. J. Silverman and A. W. Andrews, *J. Natl. Cancer Inst.*, 1977, **59**, 1557.
38. C. A. Mckillop, R. W. Owen, R. F. Bilton and E. A. Haslam, *Carcinogenesis*, 1983, **4**, 1179.
39. J. Watabe and H. Bernstein, *Mutat Res.*, 1985, **158**, 45.
40. M. Wilpart and M. Roberfroid, *Carcinogenesis*, 1986, **7**, 703.
41. A. I. Prieto, F. Ramos-Morales and J. Casadesus, *Genetics*, 2006, **174**, 575.
42. J. Wyrzykowski and M. R. Volkert, *J. Bacteriol.*, 2003, **185**, 1701.
43. G. J. S. Jenkins, F. R. D'Souza, S. H. Suzen, Z. S. Eltahir, S. A. James, J. M. Parry, P. A. Griffiths and J. N. Baxter, *Carcinogenesis*, 2007, **28**, 1.
44. A. Petitjean, E. Mathe, S. Kato, C. Ishioka, S. V. Tavtigian, P. Hainaut and M. Olivier, *Hum Mutat*, 2007, **28**, 622.
45. M. Fenech, *Mutat Res-Fund Mol. M*, 2000, **455**, 81.
46. W. S. Barnes and W. D. Powrie, *Cancer Lett.*, 1982, **15**, 317.
47. C. Fimognari, M. Nusse, R. Cesari, G. Cantelli-Forti and P. Hrelia, *Mutat Res-Genet Tox En*, 2001, **495**, 1.
48. P. Terry, J. Lagergren, W. M. Ye, O. Nyren and A. Wolk, *Int. J. Cancer*, 2000, **87**, 750.

49. A. Tzonou, L. Lipworth, A. Garidou, L. B. Signorello, P. Lagiou, C. C. Hsieh and D. Trichopoulos, *Int. J. Cancer*, 1996, **68**, 300.
50. Z. F. Zhang, R. C. Kurtz, G. P. Yu, M. Sun, N. Gargon, M. Karpeh, J. S. Fein and S. Harlap, *Nutr. Cancer*, 1997, **27**, 298.
51. J. R. Olliver, L. J. Hardie, Y. Gong, S. Dexter, D. Chalmers, K. M. Harris and C. P. Wild, *Cancer Epidemiol. Biomarkers Prev.*, 2005, **14**, 620.

CHAPTER 5
Bile Acids and Colorectal Cancer

MARK A. HULL

Section of Molecular Gastroenterology, Leeds Institute of Molecular Medicine, University of Leeds, St James's University Hospital, Leeds LS9 7TF, UK

5.1 Introduction

Secondary bile acids, particularly deoxycholic acid (DCA), are believed to play a role in human colorectal carcinogenesis based largely on the pro-tumorigenic activity of exogenous bile acids in rodent carcinogen-induced colorectal cancer (CRC) models and human observational data linking increased levels of luminal and serum bile acids with the presence of colorectal neoplasia. This chapter will summarise available evidence linking bile acids and human colorectal carcinogenesis, as well as explore current understanding of the mechanistic basis of the carcinogenic properties of bile acids.

Importantly, knowledge of intestinal bile acid transport and metabolism, coupled with increased understanding of the mechanistic basis of the pro-tumorigenic activity of bile acids against CRC cells *in vitro*, has recently led to development and testing of bile acid-based treatment and prevention strategies for sporadic and inflammatory bowel-disease-associated CRC. Existing evidence that manipulation of the luminal secondary bile acid pool and/or therapy with ursodeoxycholic acid (UDCA) may have promise for prevention of CRC will be assessed.

5.2 Multi-stage Colorectal Carcinogenesis

The model of multi-stage human colorectal carcinogenesis, based on accumulating genetic and epigenetic "hits", leading to sequential loss of

Issues in Toxicology
Bile Acids: Toxicology and Bioactivity
Edited by Gareth Jenkins and Laura J. Hardie
© Royal Society of Chemistry, 2008

tumor-suppressor gene function and oncogene activation, originally put forward by Fearon and Vogelstein[1] remains a useful conceptual tool with which to understand the slow (estimated to be 5–15 years), step-wise progression of normal colorectal mucosa through to a malignant colorectal adenocarcinoma (or cancer) *via* the classical adenoma–carcinoma sequence. However, further understanding of the molecular pathogenesis of CRC has since led to improved correlation of the clinical, morphological and molecular features of different stages of "sporadic" colorectal carcinogenesis leading to classification of separate, distinct pathways that can all lead to large-bowel malignancy *via* different, intermediate, benign lesions including the adenomatous polyp, flat adenoma and serrated hyper-plastic polyp.[2,3] For the purposes of this discussion, it is important to point out that current understanding of the role of bile acids in human colorectal carcinogenesis is based largely on rodent models of the "classical" adenoma–carcinoma pathway and *in vitro* studies of human CRC cells. The earliest identifiable, morphologically distinct lesion at early stages of human intestinal tumorigenesis is believed to be the aberrant crypt focus (ACF), which is characterised by thicker, more elongated and darkly staining crypts than neighbouring crypts.[4] ACF multiplicity has been commonly used as an end-point of tumor initiation following methylene-blue staining of the colon in rodent models of carcinogen-induced CRC (Figure 5.1), although it is still not established whether ACFs develop directly into an

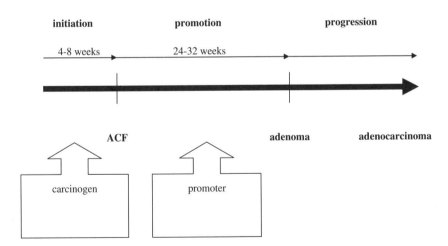

Figure 5.1 Chemical carcinogen-induced colorectal carcinogenesis in rodents. Aberrant crypt focus (ACF) multiplicity is generally regarded as an end-point for the initiation phase of carcinogenesis. Development of macroscopically visible benign adenomas and malignant adenocarcinomas (or cancers) occurs over a longer period of time during the so-called promotion phase of colorectal carcinogenesis. An agent that drives tumor initiation can be thought of as a carcinogen. An entity that does not induce tumorigenesis alone but increases tumor size and multiplicity following initiation by another carcinogen is classified as a tumor promoter.

adenoma or just represent a biomarker of future neoplastic risk in other areas
of colorectal mucosa.

5.3 Evidence that Bile Acids Play a Role in Colorectal Carcinogenesis

There are three main sources of evidence for pro-tumorigenic activity of bile
acids in the lower gastro-intestinal tract (activity in rodent CRC models,
human observational data and mechanistic studies using CRC cells *in vitro*),
which together create a strong case for a role for colorectal mucosal bile acid
exposure during human colorectal carcinogenesis.

5.3.1 Activity of Bile Acids in Rodent Models of CRC

Oral or topical administration of a number of primary (cholic acid [CA],
chenodeoxycholic acid [CDCA]) and secondary (deoxycholic acid [DCA],
lithocholic acid [LCA]) bile acids has been demonstrated to increase ACF
multiplicity in methylene-blue stained colon, or the number and size of macroscopic colorectal tumors (either adenoma and adenocarcinoma), induced by a
variety of chemical carcinogens (Table 5.1).[5–17] A consistent finding in these
studies has been that equivalent bile acid exposure to control animals not
treated with a chemical carcinogen fails to elicit tumorigenesis, implying that
bile acids act solely as a tumor promoter following tumor initiation by the
chemical carcinogen, and not as a carcinogen *per se* (Figure 5.1). However, it
has been pointed out that short-term rodent chemical carcinogenesis experiments do not mirror the much longer process of colorectal carcinogenesis
in humans, during which intestinal epithelial cell turnover, and hence DNA
replication, exceeds that of the rodent colon by approximately 30-fold.[18]
Therefore, these models may not demonstrate any relatively weak (compared
with carcinogens like AOM or MNU) direct carcinogenic properties of bile
acids. Another factor limiting the applicability of the rodent data to human
colorectal carcinogenesis is the difference in the spectrum of bile acids present in
the colonic lumen in rats (predominantly hyodeoxycholic acid) and humans
(predominantly DCA),[18] such that the mucosal response to exogenous bile
acids relevant to human carcinogenesis may differ substantially. This said, the
remarkably consistent pro-tumorigenic activity of several bile acids in a
variety of different models, is strong evidence for colorectal tumor-promoter
activity of bile acids *in vivo*.

5.3.2 Human Observational Data

Several strands of indirect evidence from human observational studies help
put the *in vivo* rodent data into the context of human colorectal carcinogenesis.

Table 5.1 Studies of the pro-tumorigenic effects of bile acids on chemical carcinogen-induced aberrant crypt focus or tumor development in the rodent colon.

Carcinogen[a]	Animal[b]	Bile acid[c]	Reference
ACF			
AOM sc	male SD rat	CDCA	Sutherland and Bird, 1994[5]
AOM sc	male F344 rat	CA/DCA/CDCA/ LCA	Seraj et al., 1997[6]
AOM sc	male SD rat	CA	Baijal et al., 1998[d7]
AOM sc	male SD rat	DCA/LCA	Baijal et al., 1998[d8]
PhIP po	male F344 rat	DCA	Ueyama et al., 2002[9]
AOM ip	male AKR/J mouse	DCA	Flynn et al., 2007[10]
Tumor			
MNNG ir	male/female CD-Fischer rat	LCA/tauro-DCA	Narisawa et al., 1974[11]
MNNG ir	female F344 rat	DCA	Reddy et al., 1976[12]
MNNG ir	female F344 rat	CA/CDCA	Reddy et al., 1977[13]
DMH sc	female Wistar rat	CA	Cruse et al., 1981[e15]
MNU ir	male F344 rat	CA	McSherry et al., 1989[16]
MNNG ir	female F344 rat	LCA	Reddy and Watanabe, 1979[14]
AOM sc	male Wistar rat	DCA	Narahara et al., 2000[17]

[a]AOM, azoxymethane; MNNG, *N*-methyl-*N'*-nitro-*N*-nitrosoguanidine; DMH, 1,2-dimethyl-hydrazine; MNU, *N*-methyl-*N*-nitrosourea; PhIP, 2-amino-1-methyl-6-phenylimidazole[4,5-b] pyridine; sc, subcutaneous; ip, intra-peritoneal; ir, intra-rectal; po, by mouth.
[b]SD, Sprague-Dawley; F, Fischer.
[c]CA, cholic acid; CDCA, chenodeoxycholic acid; DCA, deoxycholic acid; LCA, lithocholic acid.
[d]Tumor end-point *also*;
[e]Study did not demonstrate pro-tumorigenic activity of CA.

A consistent finding is that faecal secondary bile acid concentrations are higher in populations with a high overall risk of CRC compared with lower-risk groups.[19,20] This has been explained by differences in dietary fat intake,[21] with separate dietary intervention studies in healthy human volunteers demonstrating increased faecal secondary bile acid levels associated with high-fat intake.[22,23] Differences in faecal bile acid excretion between populations with a similar dietary fat and protein content has been explained by variable fibre intake.[21] One proposed mechanism of a putative beneficial effect of fibre on CRC risk is the reduction in mucosal secondary bile acid exposure due to a decrease in bile acid concentration coupled with reduced stool transit time.[24] However, a recent sub-analysis of the wheat bran fibre colon polyp trial failed to demonstrate any significant effect of high-fibre dietary intervention for a median of 2.4 years on faecal water DCA and LCA concentrations in patients aged 40–80 years, who had previously had at least one colorectal adenoma removed.[25] By contrast, a short-term (10 week) high-fibre diet has been demonstrated to reduce stool secondary bile acid concentrations (but not excretion) in healthy volunteers.[26] Whether discrepancies between these studies are due to methodological differences in stool bile acid measurement or represent real differences in the effect of fibre intake in different human populations await clarification.

It is well recognised that the faecal bile acid content of random stool samples is highly variable with marked daily variation.[27] Therefore, studies testing the association between luminal bile acid exposure and the presence of colorectal neoplasia have usually measured serum bile acid levels, which demonstrate less variability and are believed to reflect the total bile acid pool more accurately.[28,29] Serum DCA levels have been shown to be higher in individuals with a colorectal adenoma compared with individuals without a neoplasm.[30,31] Only one study has assessed future risk of CRC in a prospective study of serum bile-acid levels.[32] The study was hampered by the small sample size (46 CRC cases). There were no significant differences in the absolute concentrations of primary and secondary bile acids or DCA/CA ratio between cases and controls although there was a trend towards increased CRC risk for those with a DCA/CA ratio in the top third of values (relative risk 3.9 [95% confidence interval 0.9–17.0; $p = 0.1$]).[32] It will be important to test the possible utility of the DCA/CA ratio as a CRC risk biomarker in larger, adequately powered studies. A recent study has demonstrated increased levels of allo-DCA and allo-LCA metabolites in the stool of CRC patients compared with healthy controls.[33]

Additionally, two studies have measured colorectal epithelial cell proliferation and apoptosis in human non-neoplastic mucosa in combination with serum bile acid quantification. Ochsenkuhn *et al.*[34] have reported a positive correlation between serum DCA levels and proliferation measured by flow cytometric cell cycle analysis. However, a more recent study of colorectal adenoma patients failed to detect a correlation between serum DCA and immunohistochemical Ki-67 antigen labelling.[35] Instead, this latter study revealed a positive correlation between serum DCA and the degree of TUNEL-positive epithelial cell apoptosis.[35]

Two clinical scenarios also allow testing of the hypothesis that increased colonic mucosal bile acid exposure drives human colorectal carcinogenesis. Firstly, cholecystectomy leads to increased luminal secondary bile acid production.[36] However, an association between cholecystectomy and increased overall CRC risk is not well established with marked heterogeneity between a large number of (mainly case-control) studies, in which the CRC risk is lower in population-based rather than hospital-based designs, and conflicting conclusions from separate meta-analyses.[37,38] By contrast, cohort studies have generally failed to detect an increase in CRC risk following cholecystectomy, but this may be due to insufficient follow-up duration. In the largest cohort study published, there was an increased risk of proximal, but not distal colon cancer, consistent with the idea that secondary bile acid exposure is likely to be highest in the proximal colon.[39] The risk of proximal small intestinal cancer is also increased after cholecystectomy.[39]

Secondly, ileal resection disrupts enterohepatic circulation of bile acids due to loss of the ileal sodium-dependent bile acid transporter (ISBT) leading to increased proximal colonic exposure to primary bile acids. In the 1970s, it was noted that ileal resection promoted DMH-induced colorectal carcinogenesis.[40] More recently, ileal, but not jejunal, resection has been demonstrated to increase faecal bile acid concentration and AOM-induced colonic tumor multiplicity in rats who received oral DCA.[41] Surprisingly, there are no published data on CRC risk following ileal resection in humans, perhaps explained by the difficulty in distinguishing the effect of ileal resection from that of co-existent colitis in patients with Crohn's disease who have undergone surgery for an ileitis. It would be interesting to investigate neoplastic risk in those individuals with post-infective or idiopathic bile salt malabsorption. One group have studied the association between single nucleotide polymorphisms (SNPs) in the human ISBT gene (*SLC10A2*) and risk of colorectal adenoma.[42] There was a significant 2-fold higher risk of adenoma associated with the 169C→T SNP.[42] However, this SNP is predicted to be non-functional, suggesting linkage dis-equilibrium with an unidentified at-risk *SLC10A2* allele.

In summary, despite methodological difficulties measuring mucosal bile acid exposure in humans, diverse sets of observational data are consistent with the notion that increased colonic bile acid levels pre-dispose to colorectal neoplasia. Combined with the rodent data on the effects of exogenous bile acid administration on chemical carcinogenesis, this indirect evidence is a strong indicator of a role for mucosal bile acid exposure during human colorectal carcinogenesis. However, what remains unclear is the relative contribution of bile acids to colorectal carcinogenesis compared with other known environmental/behavioural aetiological factors (*e.g.* obesity, low fruit and vegetable intake, *etc.*).

5.4 Mechanisms of the Carcinogenic Activity of Secondary Bile Acids

A large body of evidence has accumulated describing multiple mechanisms of the carcinogenic activity of bile acids. This work has been summarised recently in a detailed review by Bernstein and colleagues.[18]

Bile acids are believed to be mutagenic, despite the lack of activity of bile acids in the Ames *Salmonella* mutagen assay, which may be explained by the relative insensitivity of this particular detection system. Bile acids drive generation of reactive oxygen species (ROS) and reactive nitrogen species (RNS) such as the superoxide radical (O_2^-) by multiple mechanisms, which are likely to include induction of cyclo-oxygenase (COX)[43] and lipoxygenase activity,[44] as well as up-regulation of nitric oxide synthase (NOS) isoforms, particularly NOS2.[45] ROS and RNS induce oxidative DNA damage leading to DNA mutations (reviewed in detail in reference 18). Whether bile acid mutagenesis is relevant only for tumor initiation or is also relevant to the promotion phase of colorectal carcinogenesis is not firmly established. As secondary bile acids appear to have tumor-promoter (but no carcinogenic) activity in the rodent models outlined in Section 5.3.1, it is important to note that DCA has been demonstrated to increase the incidence of oncogenic *K-ras* mutations (a common genetic event associated with promotion and progression phases of colorectal carcinogenesis, rather than the initiation phase) in AOM-induced colonic tumors.[17]

Deoxycholic acid induces caspase-dependent colonocyte apoptosis at physiologically relevant concentrations.[18] There is evidence that oxidative DNA damage is likely to contribute to the pro-apoptotic activity of this secondary bile acid.[43] Other postulated mechanisms include mitochondrial stress[46] and alteration of plasma membrane structure leading to receptor tyrosine kinase activation and ceramide release.[47] Several other signal transduction pathways have been implicated in DCA-induced apoptosis, including the protein kinase C pathway and p38 mitogen-activated protein kinase signalling.[43] It has been proposed that long-term exposure of intestinal epithelial cells to bile acids leads to selection of cells resistant to apoptosis, which could then accelerate carcinogenesis due to increased accumulation of oncogenic mutations.[18] There are some data from studies of low concentrations of DCA on human colonocytes *in vitro*[48] and the effect of chronic CA administration on AOM-induced intestinal epithelial cell apoptosis in rats,[49] which support this idea. The mechanistic basis of acquired apoptosis resistance in intestinal epithelial cells appears to involve nuclear factor κB activation.[46,50] However, in the only published human observational study of its kind, increased serum DCA levels (taken to represent higher long-term mucosal secondary bile acid exposure) correlated with increased epithelial cell apoptosis measured by the TUNEL technique in human colorectal mucosa,[35] questioning the *in vivo* relevance of acquired apoptosis resistance to bile acid-induced colorectal carcinogenesis.

Deoxycholic acid has also been demonstrated to activate other pivotal oncogenic pathways in CRC cells *in vitro* including β-catenin/T-cell factor-mediated transcription, extra-cellular signal-regulated kinase activation downstream of the epidermal growth factor receptor and Jun-N-terminal kinase activation.[18,51] It is currently not known which of the many cellular consequences of exogenous DCA exposure represent primary or direct effects of DCA, as opposed to secondary, indirect phenomena.

5.5 Colorectal Cancer Chemoprevention Strategies Targeting Bile Acids (Illustrated in Figure 5.2)

5.5.1 Ursodeoxycholic Acid

In contrast to the effects of secondary bile acids like DCA, UDCA, which is found at a high concentration in bear bile but only in trace amounts in humans, has anti-neoplastic activity *in vitro* and *in vivo*. UDCA has significant chemopreventative activity in rodent models of "sporadic"[52] and "colitis-associated"[53,54] colorectal carcinogenesis induced by chemical carcinogens. UDCA has been used for treatment of patients with primary sclerosing

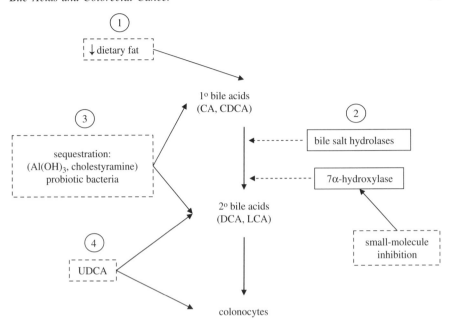

Figure 5.2 Therapeutic interventions for decreasing colorectal mucosal bile acid exposure as a CRC chemoprevention strategy. 1) Lifestyle modifications including reduction in dietary animal fat and increased fibre intake may, at least partly, be explained by reduction in luminal primary (cholic acid [CA] and chenodeoxycholic acid [CDCA]) and secondary (deoxycholic acid [DCA] and lithocholic acid [LCA]) bile acids. 2) Reduction of secondary bile acids, which are believed to have pro-carcinogenic activity could be obtained by decreased bacterial conversion from primary bile acids. 3) Alternatively, bile acids could be sequestered by chemical binding agents, *e.g.* aluminium hydroxide (Al(OH)$_3$) or probiotic bacteria. 4) Exogenous ursodeoxycholic acid (UDCA) can reduce the luminal proportion of secondary bile acids and also has direct anti-neoplastic activity on colonocytes *in vitro*.

cholangitis, the majority of whom have ulcerative colitis and elevated "colitis-associated" CRC risk. Several studies have investigated whether UDCA use decreases dysplasia and CRC risk in this context. There have been three, small studies, of which two have demonstrated decreased neoplastic risk in the UDCA arm compared with no treatment.[55,56] The other published (historical cohort) study failed to reveal any difference related to UDCA use.[57] These studies and one report of a reduction in colorectal adenoma recurrence in primary biliary cirrhosis patients taking UDCA[58] have prompted the evaluation of UDCA 8–10 mg/kg daily for secondary prevention of "sporadic" colorectal adenomas in a large (1285 individuals) randomised, placebo-controlled study.[59] This demonstrated a nonsignificant 12% reduction in adenoma recurrence in the UDCA arm, but a significant reduction in recurrence of adenomas with high-grade dysplasia associated with UDCA use.[59]

The study authors pointed out the relatively short duration of this study may have limited the magnitude of any UDCA effect observed, particularly as the beneficial effect of UDCA on dysplasia incidence in one of the primary sclerosing cholangitis studies only became apparent after longer duration therapy.[56]

The mechanistic basis of the anti-neoplastic activity of UDCA and the explanation for the significant difference in bioactivity of UDCA compared with DCA despite marked similarity in chemical structure remain unresolved. UDCA administration in healthy volunteers and colorectal adenoma patients has been demonstrated to decrease the proportion of DCA in aqueous phase stool.[59,60] Therefore, one possible mechanism of the chemopreventative activity of UDCA is reduction of mucosal secondary bile acid exposure. Consistent with this idea, UDCA administration has been demonstrated to reduce the incidence of *K-ras* mutations and decrease Cox-2 expression in AOM-induced tumors,[61] which is the opposite of the reported effects of DCA in the same model.[17] However, it is clear that exogenous administration of UDCA has direct anti-neoplastic activity on human CRC cells *in vitro*, either alone or in combination with DCA, including anti-proliferative and anti-apoptotic effects, as well as induction of cell senescence.[62–64]

5.5.2   Other Agents Targeting Mucosal Bile Acid Exposure

Other agents that bind luminal bile acids and reduce mucosal bile acid exposure would be expected to have anti-neoplastic activity if bile acids drive colorectal carcinogenesis. A recent study has demonstrated that administration of the bile salt-binding anion-exchange resin cholestyramine decreases rat colonocyte apoptosis frequency compared with control animals, consistent with the concept of antagonism of the effect of secondary bile acids on intestinal epithelial cells.[65] However, a study in the early 1970s reported that cholestyramine administration actually increased carcinogen-induced colorectal tumor multiplicity[66] and another more recent study demonstrated no effect of cholestyramine on DMH-induced colorectal carcinogenesis in the presence of cholic acid.[15] It would be very interesting to determine the effect of this drug on AOM-induced colorectal carcinogenesis combined with DCA administration. Unfortunately, clinical use of cholestyramine is limited by poor oral tolerability and pharmacodynamic interactions with multiple drugs.

Aluminium hydroxide, which has been used widely as an antacid, also binds bile acids. However, there has been no clinical indication of altered CRC risk associated with its use and there was no observable difference in DMH-induced colorectal tumor incidence in rats receiving oral aluminium hydroxide compared with sham treatment.[15]

Alternative potential strategies for reduction of mucosal secondary bile acid exposure are to target deconjugation of glycine/taurine bile salts by bacterial bile salt hydrolases and/or bacterial 7-dehydroxylation of primary bile acids to secondary bile acids. Sequestration of bile acids in the intestinal lumen using probiotic bacteria has also been proposed as an area for future research.[51]

5.5.3 Therapeutic Anti-cancer Use of Bile Acid Conjugates

Recently, the therapeutic utility of bile acids as hydrophobic "carrier" molecules for other pharmaceutical agents has been recognised.[67] Bile acid derivatives may also utilise specific bile acid transporter proteins, which are expressed in human colorectal adenoma and CRC tissue, for efficient intra-cellular drug targeting, particularly for treatment of established CRC rather than chemoprevention at earlier stages of colorectal carcinogenesis.[68] For example, the cisplatin conjugate *cis*-diammine-bisursodeoxycholate-platinum(II) (Bamet-UD2) has been demonstrated to induce apoptosis of LS174T human CRC cells and overcome cisplatin resistance of a resistant LS174T cell clone *in vitro*.[69] Other bile acid derivatives including a folate-heparin-lithocholate conjugate and the glycine methyl ester conjugate of UDCA have anti-angiogenic properties and anti-tumor activity *in vivo*.[70,71]

5.6 Summary

The combination of consistent pro-tumorigenic activity of exogenous bile acids in many independent studies using rodent chemical carcinogen-induced CRC models and the human observational data linking mucosal bile acid exposure with colorectal neoplasia together make a convincing case for a role for bile acids (particularly the secondary bile acid DCA) in human colorectal carcinogenesis. However, it is still not possible to determine the relative magnitude of the effect of bile acids compared with the other genetic and environmental factors, which drive colorectal carcinogenesis, so that the overall relevance (and hence importance as a target for prevention and treatment of CRC) of bile acids is impossible to gauge.

High-throughput assay of bile acids is technically demanding and expensive. Combined with high intra- and inter-individual variability in stool bile acid levels, it is unlikely that bile acid measurement will find a role as a "biomarker" of future colorectal neoplastic risk.

Anti-neoplastic activity of UDCA was demonstrated first in the context of ulcerative colitis-associated colorectal carcinogenesis. Subsequently, encouraging (but not definitive) results have been obtained in clinical trials of UDCA for prevention of "sporadic" colorectal adenoma recurrence, which should prompt further evaluation of UDCA for polyp prevention, particularly given its excellent safety profile compared with other candidate chemoprevention agents such as the nonsteroidal anti-inflammatory drugs.

The development of lipophillic bile acid-drug conjugates and other bile acid derivatives is an important advance with exciting therapeutic possibilities.

Acknowledgement

Work in the Hull Laboratory is supported by the Medical Research Council (UK), Yorkshire Cancer Research, Cancer Research UK and World Cancer Research Fund International.

References

1. E. R. Fearon and B. Vogelstein, A genetic model for colorectal tumourigenesis, *Cell*, 1990, **61**, 759.
2. J. R. Jass, Classification of colorectal cancer based on correlation of clinical, morphological and molecular features, *Histopathology*, 2007, **50**, 113.
3. C. N. Arnold, A. Goel, H. E. Blum and C. R. Boland, Molecular pathogenesis of colorectal cancer, *Cancer*, 2005, **104**, 2035.
4. D. P. Hurlstone and S. S. Cross, Role of aberrant crypt foci detected using high-magnification-chromoscopic colonoscopy in human colorectal carcinogenesis, *J. Gastroenterol. Hepatol.*, 2005, **20**, 173.
5. L. A. M. Sutherland and R. P. Bird, The effect of chenodeoxycholic acid on the development of aberrant crypt foci in the rat colon, *Cancer Lett.*, 1994, **76**, 101.
6. M. J. Seraj, A. Umemoto, A. Kajikawa, S. Mimura, T. Kinouchi, Y. Ohnishi and Y. Monden, Effects of dietary bile acids on formation of azoxymethane-induced aberrant crypt foci in F344 rats, *Cancer Lett.*, 1997, **115**, 97.
7. P. K. Baijal, E. P. Clow, D. W. Fitzpatrick and R. P. Bird, Tumour-enhancing effects of cholic acid are exerted on the early stages of colon carcinogenesis *via* induction of aberrant crypt foci with enhanced growth phenotype, *Can. J. Physiol. Pharmacol.*, 1998, **76**, 1095.
8. P. K. Baijal, D. W. Fitzpatrick and R. P. Bird, Comparative effects of secondary bile acids, deoxycholic and lithocholic acids, on aberrant crypt foci growth in the postinitiation phases of colon carcinogenesis, *Nutr. Cancer*, 1998, **31**, 81.
9. Y. Ueyama, Y. Monden, X. B. He, C. X. Lin, M. A. Momen, S. Mimura and A. Umemoto, Effects of bile acids on 2-amino-1-methyl-6-phenylimidazo-[4,5-b]pyridine-induced aberrant crypt foci and DNA adduct formation in the rat colon, *J. Exp. Clin. Cancer Res.*, 2002, **21**, 577.
10. C. Flynn, D. C. Montrose, D. L. Swank, M. Nakanishi, J. N. M. Ilsley and D. W. Rosenberg, Deoxycholic acid promotes the growth of colonic aberrant crypt foci, *Mol. Carcinog.*, 2007, **46**, 60.
11. T. Narisawa, N. E. Magadia, J. H. Weisburger and E. L. Wynder, Promoting effect of bile acids on colon carcinogenesis after intra-rectal instillation of N-methyl-N'-nitro-N-nitrosoguanidine in rats, *J. Natl. Cancer Inst.*, 1974, **53**, 1093.
12. B. S. Reddy, T. Narasawa, J. H. Weisburger and E. L. Wynder, Promoting effect of sodium deoxycholate on colon adenocarcinomas in germ-free rats, *J. Natl. Cancer Inst.*, 1976, **56**, 441.
13. B. S. Reddy, K. Watanabe, J. H. Weisburger and E. L. Wynder, Promoting effect of bile acids in colon carcinogenesis in germ-free and conventional F344 rats, *Cancer Res.*, 1977, **37**, 3238.
14. B. S. Reddy and K. Watanabe, Effect of cholesterol metabolites and promoting effect of lithocholic acid in colon carcinogenesis in germ-free and conventional F344 rats, *Cancer Res.*, 1979, **39**, 1521.

15. J. P. Cruse, M. R. Lewin and C. G. Clark, The effects of cholic acid and bile salt binding agents on 1,2-dimethylhydrazine-induced colon carcinogenesis in the rat, *Carcinogenesis*, 1981, **2**, 439.
16. C. K. McSherry, B. I. Cohen, V. D. Bokkenheuser, E. H. Mosbach, J. Winter, N. Matoba and J. Scholes, Effects of calcium and bile-acid feeding on colon tumours in the rat, *Cancer Res.*, 1989, **49**, 6039.
17. H. Narahara, M. Tatsuta, H. Iishi, M. Baba, N. Uedo, N. Sakai, H. Yano and S. Ishiguro, K-*ras* point mutation is associated with enhancement by deoxycholic acid of colon carcinogenesis induced by azoxymethane, but not with its attenuation by all-*trans*-retinoic acid, *Int. J. Cancer*, 2000, **88**, 157.
18. H. Bernstein, C. Bernstein, C. M. Payne, K. Dvorakova and H. Garewal, Bile acids as carcinogens in human gastro-intestinal cancers, *Mutation Res.*, 2005, **589**, 47.
19. B. S. Reddy and E. L. Wynder, Large bowel carcinogenesis: Fecal constituents of populations with diverse incidence rates of colon cancer, *J. Natl. Cancer Inst.*, 1973, **50**, 1437.
20. T. Kishida, F. Taguchi, L. Feng, A. Tatsuguchi, J. Sato, S. Fujimori, H. Tachikawa, Y. Tamagawa, Y. Yoshida and M. Kobayashi, Analysis of bile acids in colon residual liquid or fecal material in patients with colorectal neoplasia and control subjects, *J. Gastroenterol.*, 1997, **32**, 306.
21. B. S. Reddy, Diet and excretion of bile acids, *Cancer Res.*, 1981, **41**, 3766.
22. B. S. Reddy, J. H. Weisburger and E. L. Wynder, Effect of high-risk and low-risk diets for colon carcinogenesis on fecal microflora and steroids in man, *J. Nutr.*, 1975, **105**, 878.
23. J. H. Cummings, H. S. Wiggins, D. J. A. Jenkins, H. Houston, T. Jivraj, B. S. Drasar and M. J. Hill, Influence of diets high and low in animal fat on bowel habit, gastro-intestinal transit time, fecal microflora, bile acid, and fat excretion, *J. Clin. Invest.*, 1978, **61**, 953.
24. J. Salter, M. Chaplin, J. Dickerson and J. Davies, Bile acids and health: Is fibre the answer? *Nutr. Food Sci.*, 1996, **6**, 29.
25. D. S. Alberts, J. G. Einspahr, D. L. Earnest, M. F. Krutzsch, P. Lin, L. M. Hess, D. K. Heddens, D. J. Roe, M. E. Martinez, G. Salen and A. K. Batta, Fecal bile-acid concentrations in a subpopulation of the wheat bran fiber colon polyp trial, *Cancer Epidemiol. Biomarkers Prev.*, 2003, **12**, 197.
26. F. M. Nagengast, G. van den Ban, J. P. Ploeman, R. Leenen, P. L. Zock, M. B. Katan, M. P. Hectors, A. F. de Haan and J. H. Tongeren, The effect of a natural high-fibre diet on faecal and biliary bile acids, faecal pH and whole-gut transit time in man. A controlled study, *Eur. J. Clin. Nutr.*, 1993, **47**, 631.
27. K. D. R. Setchell, J. A. Ives, G. C. Cashmore and A. M. Lawson, On the homogeneity of stools with respect to bile-acid composition and normal day-to-day variations: A detailed qualitative and quantitative study using capillary column gas chromatography-mass spectrometry, *Clin. Chim. Acta*, 1987, **162**, 257.
28. S. E. McGarr, J. M. Ridlon and P. B. Hylemon, Diet, anaerobic metabolism, and colon cancer, *J. Clin. Gastroenterol.*, 2005, **39**, 98.

29. E. Bayerdorffer, G. A. Mannes, T. Ochsenkuhn, P. Dirschedl and G. Paumgartner, Variation of serum bile acids in patients with colorectal adenomas during a one-year follow-up, *Digestion*, 1994, **55**, 121.

30. E. Bayerdorffer, G. A. Mannes, W. O. Richter, Ochsenkuhn, B. Wiebecke, W. Kopcke and G. Paumgartner, Increased serum deoxychloic acid levels in men with colorectal adenomas, *Gastroenterology*, 1993, **104**, 145.

31. E. Bayerdorffer, G. A. Mannes, T. Ochsenkuhn, P. Dirschedl, B. Wiebecke and G. Paumgartner, Unconjugated secondary bile acids in the serum of patients with colorectal adenomas, *Gut*, 1995, **36**, 268.

32. V. Costarelli, T. J. Key, P. N. Appleby, D. S. Allen, I. S. Fentiman and T. A. B. Sanders, A prospective study of serum bile-acid concentrations and colorectal cancer risk in post-menopausal women on the island of Guernsey, *Br. J. Cancer*, 2002, **86**, 1741.

33. T. Tadano, M. Kanoh, M. Matsumoto, K. Sakamoto and T. Kamano, Studies of serum and feces bile acids determination by gas chromatography-mass spectrometry, *Rinsho Byori*, 2006, **54**, 103.

34. T. Ochsenkuhn, E. Bayerdorffer, A. Meining, M. Schinkel, C. Thiede, V. Nussler, M. Sackmann, R. Hatz, A. Neubauer and G. Paumgartner, Colonic mucosal proliferation is related to serum deoxycholic acid levels, *Cancer*, 1999, **85**, 1664.

35. M. Fracchia, G. Galatola, I. Sarotto, V. Guraldo, M. Perona, A. Pera and M. Risio, Serum bile acids, programmed cell death and cell proliferation in the mucosa of patients with colorectal adenomas, *Dig. Liver Dis.*, 2005, **37**, 509.

36. G. W. Hepner, A. F. Hofmann, J. R. Malagelada, P. A. Szczepanik and P. D. Klein, Increased bacterial degradation of bile acids in cholecystectomized patients, *Gastroenterology*, 1974, **66**, 556.

37. E. Giovannucci, G. A. Colditz and M. J. Y. Stampfer, A meta-analysis of cholecystectomy and risk of colorectal cancer, *Gastroenterology*, 1993, **105**, 130.

38. F. D. Reid, P. M. Mercer, M. Harrison and T. Bates, Cholecystectomy as a risk factor for colorectal cancer: A meta-analysis, *Scand. J. Gastroenterol.*, 1996, **31**, 160.

39. J. Lagergren, W. Ye and A. Ekbom, Intestinal cancer after cholecystectomy: Is bile involved in carcinogenesis? *Gastroenterology*, 2001, **121**, 542.

40. J. E. Oscarson, H. F. Veen, J. S. Ross and R. A. Malt, Ileal resection potentiates 1,2-dimethylhydrazine-induced colonic carcinogenesis, *Ann. Surg.*, 1979, **189**, 503.

41. R. Kanamoto, N. Azuma, H. Suda, T. Seki, Y. Tsuchihashi and K. Iwami, Elimination of Na^+-dependent bile-acid transporter from small intestine by ileum resection increase colonic tumourigenesis in the rat fed deoxycholic acid, *Cancer Lett.*, 1999, **145**, 115.

42. W. Wang, S. Xue, S. A. Ingles, Q. Chen, A. T. Diep, H. D. Frankl, A. Stolz and R. W. Haile, An association between genetic polymorphisms in the ileal sodium-dependent bile-acid transporter gene and the risk of colorectal adenomas, *Cancer Epidemiol. Biomarkers Prev.*, 2001, **10**, 931.

43. B. Glinghammar, H. Inoue and J. J. Rafter, Deoxycholic acid causes DNA damage in colonic cells with subsequent induction of caspases, *COX-2* promoter activity and the transcription factors NF-kB and AP-1, *Carcinogenesis*, 2002, **23**, 839.
44. F. R. DeRubertis, P. A. Craven and R. Saito, Bile salt stimulation of colonic epithelial cell proliferation. Evidence for involvement of lipoxygenase products, *J. Clin. Invest.*, 1984, **74**, 1614.
45. Y. Hirose, C. V. Rao and B. S. Reddy, Modulation of inducible nitric oxide synthase expression in rat intestinal epithelial cells by colon tumour promoters, *Int. J. Oncol.*, 2001, **18**, 141.
46. C. M. Payne, C. Weber, C. Crowley-Skillicorn, K. Dvorak, H. Bernstein, C. Bernstein, H. Holubec, B. Dvorakova and H. Garewal, Deoxycholate induces mitochondrial oxidative stress and activates NF-κB through multiple mechanisms in HCT-116 colon epithelial cells, *Carcinogenesis*, 2007, **28**, 215.
47. S. Jean-Louis, S. Akare, M. A. Ali, E. A. Mash, E. Meuillet and J. D. Martinez, Deoxycholic acid induces intra-cellular signalling through membrane perturbation, *J. Biol. Chem.*, 2006, **281**, 14948.
48. C. L. Crowley-Weber, C. M. Payne, M. Gleason-Guzman, G. S. Watts, B. Futscher, C. N. Waltmire, C. Crowley, K. Dvorakova, C. Bernstein, M. Craven, H. Garewal and H. Bernstein, Development and molecular characterization of HCT-116 cell lines resistant to the tumour promoter and multiple stress-inducer deoxycholate, *Carcinogenesis*, 2002, **23**, 2063.
49. B. A. Magnuson, N. Shirtliff and R. P. Bird, Resistance of aberrant crypt foci to apoptosis induced by azoxymethane in rats chronically fed cholic acid, *Carcinogenesis*, 1994, **15**, 1459.
50. D. J. Turner, S. M. Alaish, T. Zou, J. N. Rao, J.-Y. Wang and E. D. Strauch, Bile salts induce resistance to apoptosis through NF-kB-mediated XIAP expression, *Ann. Surg.*, 2007, **245**, 415.
51. J. M. Ridlon, D.-J. Kang and P. B. Hylemon, Bile salt transformations by human intestinal bacteria, *J. Lipid Res.*, 2006, **47**, 241.
52. R. K. Wali, D. Stoiber, L. Nguyen, J. Hart, M. D. Sitrin, T. Brasitus and M. Bissonnette, Ursodeoxycholic acid inhibits the initiation and post-initiation phases of azoxymethane-induced colonic tumour development, *Cancer Epidemiol. Biomarkers Prev.*, 2002, **11**, 1316.
53. C. Loddenkemper, S. Keller, M.-L. Hanski, M. Cao, G. Jahreis, H. Stein, M. Zeitz and C. Hanski, Prevention of colitis-associated carcinogenesis in a mouse model by diet supplementation with ursodeoxycholic acid, *Int. J. Cancer*, 2006, **118**, 2750.
54. H. Kohno, R. Suzuki, Y. Yasui, S. Miyamoto, K. Wakabayashi and T. Tanaka, Ursodeoxycholic acid versus sulfasalazine in colitis-related colon carcinogenesis in mice, *Clin. Cancer Res.*, 2007, **13**, 2519.
55. B. Y. Tung, M. J. Emond, R. C. Haggitt, M. P. Bronner, M. B. Kimmey, K. V. Kowdley and T. A. Brentnall, Ursodiol use is associated with lower prevalence of colonic neoplasia in patients with ulcerative colitis and primary sclerosing cholangitis, *Ann. Intern. Med.*, 2001, **134**, 89.

56. D. S. Pardi, E. V. Loftus, W. K. Kremers, J. Keach and K. Lindor, Ursodeoxycholic acid acts as a chemopreventive agent in patients with primary sclerosing cholangitis, *Gastroenterology*, 2001, **121**, 900.

57. J. M. Wolf, L. A. Rybicki and B. A. Lashner, The impact of ursodeoxycholic acid on cancer, dysplasia and mortality in ulcerative colitis patients with primary sclerosing cholangitis, *Aliment. Pharmacol. Ther.*, 2005, **22**, 783.

58. L. Serfaty, A. De Leusse, O. Rosmorduc, B. Desaint, J.-F. Flejou, O. Chaouilleres, R. E. Poupon and R. Poupon, Ursodeoxycholic acid therapy and the risk of colorectal adenoma inpatients with primary biliary cirrhosis: An observational study, *Hepatology*, 2003, **38**, 203.

59. D. S. Alberts, M. E. Martinez, L. M. Hess, J. G. Einspahr, S. B. Green, A. K. Bhattacharyya, J. Guillen, M. Krutzsch, A. K. Batta, G. Salen, L. Fales, K. Koonce, D. Parish, M. Clouser, D. Roe and P. Lance, Phase III trial of ursodeoxycholic acid to prevent colorectal adenoma recurrence, *J. Natl. Cancer Inst.*, 2005, **97**, 846.

60. L. M. Hess, M. F. Krutzsch, J. Guillen, H. H. Chow, J. Einspahr, A. K. Batta, G. Salen. M. E. Reid, D. L. Earnest and D. S. Alberts, Results of a phase I multiple-dose clinical study of ursodeoxycholic acid, *Cancer Epidemiol. Biomarkers Prev.*, 2004, **13**, 861.

61. S. Khare, S. Cerda, R. K. Wali, F. C. von Lintig, M. Tretiakova, L. Joseph, D. Stoiber, G. Cohen, K. Nimmagadda, J. Hart, M. D. Sitrin, G. R. Boss and M. Bissonnette, Ursodeoxycholic acid inhibits Ras mutations, wild-type Ras activation, and cyclo-oxygenase-2 expression in colon cancer, *Cancer Res.*, 2003, **63**, 3517.

62. J. D. Martinez, E. D. Stratagoules, J. M. LaRue, A. A. Powell, P. R. Gause, M. T. Craven, C. M. Payne, M. B. Powell, E. W. Gerner and D. L. Earnest, Different bile acids exhibit distinct biological effects: The tumour promoter deoxycholic acid induces apoptosis and the chemopreventive agent ursodeoxycholic acid inhibits cell proliferation, *Nutr. Cancer*, 1998, **31**, 111.

63. E. Im and J. D. Martinez, Ursodeoxycholic acid (UDCA) can inhibit deoxycholic acid (DCA)-induced apoptosis *via* modulation of EGFR/ERK signalling in human colon cancer cells, *J. Nutr.*, 2004, **134**, 483.

64. S. Akare, S. Jean-Louis, W. Chen, D. J. Wood, A. A. Powell and J. D. Martinez, Ursodeoxycholic acid modulates histone acetylation and induces differentiation and senescence, *Int. J. Cancer*, 2006, **119**, 2958.

65. L. Lack, H. B. Suliman, A. A. Rahman and M. B. Abou-Donia, Cholestyramine feeding lowers number of colonic apoptotic cells in rat, *J. Toxicol. Environ. Health*, 2005, **68**, 1963.

66. N. D. Nigro, N. Bhadrachari and C. Chomchai, A rat model for studying colonic cancer: Effect of cholestyramine on induced tumours, *Dis. Colon Rectum*, 1973, **16**, 438.

67. M. Mikov, J. P. Fawcett, K. Kuhajda and S. Kevresan, Pharmacology of bile acids and their derivatives: Absorption promoters and therapeutic agents, *Eur. J. Drug Metab. Pharmacokinet.*, 2006, **31**, 237.

68. M. R. Ballestero, M. J. Monte, O. Briz, F. Jimenez, F. Gonzalez-San Martin and J. J. G. Marin, Expression of transporters potentially involved in the targeting of cytostatic bile-acid derivatives to colon cancer and polyps, *Biochem. Pharmacol.*, 2006, **72**, 729.
69. M. J. Monte, M. R. Ballestero, O. Briz, M. J. Perez and J. J. G. Marin, Proapoptotic effect on normal and tumour intestinal cells of cytostatic drugs with enterohepatic organotropism, *J. Pharmacol. Exp. Ther.*, 2005, **315**, 24.
70. M. K. Yu, D. Y. Lee, Y. S. Kim, K. Park, S. A. Park, D. H. Son, G. Y. Lee, J. H. Nam, S. Y. Kim, I. S. Kim, R. W. Park and Y. Byun, Anti-angiogenic and apoptotic properties of a novel amphiphilic folate-heparin-lithocholate derivative having cellular internality for cancer therapy, *Pharm. Res.*, 2007, **24**, 705.
71. H. Suh, E.-J. Jung, T. H. Kim, H. Y. Lee, Y. H. Park and K. W. Kim, Anti-angiogenic activity of ursodeoxycholic acid and its derivatives, *Cancer Lett.*, 1997, **113**, 117.

CHAPTER 6
Bile Acids and Oesophageal Adenocarcinoma (OA)

GARETH JENKINS[*] AND JAMES CRONIN

Institute of Life Science, Swansea School of Medicine, Swansea University, Swansea SA28PP, UK

6.1 Oesophageal Adenocarcinoma

The incidence rates of oesophageal adenocarcinoma (OA) have risen rapidly in Western industrialised countries in the last quarter of a century.[1] OA is caused by chronic gastro-oesophageal reflux disease (GORD). GORD effects up to 20% of the general population and around 10% of these (*i.e.* 1–2% of the general population) acquire a metaplastic condition known as Barrett's oesophagus.[2-4] In Barrett's patients, the normal squamous lining of the lower tubular oesophagus is replaced (after chronic GORD) by intestinal columnar epithelium, reminiscent of intestinal tissue. For some reason, this metaplastic tissue produces an unstable phenotype, with neoplastic progression occurring in 10% of patients, leading to OA, a form of cancer that is associated with survival rates <10% at 5 years (CR-UK figures: http://info.cancerresearchuk. org/cancerstats/survival/latestrates).

In GORD, the refluxate is thought to be responsible for directly damaging the squamous mucosa, resulting in an inflammatory response and the triggering of the development of a more resistant, mucin-secreting intestinal tissue type, derived from squamous stem cells.[4] Sustained GORD is then responsible for the metaplasia to adenocarcinoma progression. Initially, it was thought that acid reflux played the most significant role in the development of BO and its

[*]corresponding author

Issues in Toxicology
Bile Acids: Toxicology and Bioactivity
Edited by Gareth Jenkins and Laura J. Hardie
© Royal Society of Chemistry, 2008

sequelae adenocarcinoma. This led to the development of acid-suppressing drugs to minimise acid reflux with a view to eliminating symptoms and preventing carcinogenesis in the lower oesophagus of GORD patients. Whilst acid-suppression therapies are extremely effective at managing symptoms of GORD, there is little evidence that they have had an impact on carcinogenesis in GORD patients. Indeed, despite their introduction in the 1980s (H_2 receptor antagonists, followed by proton pump inhibitors in the 1990s), there has been an undented rise in the incidence of OA in Western countries over the past 3 decades.[5,6] This has led to the theory that some nonacid constituent of reflux may be driving carcinogenesis. It is also possible that some other changing lifestyle trend is implicated in the rising incidence of OA, notably obesity. It has been shown recently in a study of over a million women in the UK, that body mass index (BMI) is significantly associated with OA development.[7] Nonetheless, there is marked current interest in bile acids as oesophageal carcinogens. Here, we detail the potential contribution of bile acids to oesophageal carcinogenesis. Whilst acid reflux undoubtedly plays a part in this process, through inducing mucosal damage and compensatory proliferation, stimulation of signalling pathways, *etc.*, we concentrate here solely on the roles for bile acids.

6.2 Oesophageal Bile Acid Exposure

After their synthesis in hepatocytes, bile acids are excreted as amides conjugated with glycine or taurine.[8,9] This amidation increases aqueous solubility at acidic pH, increases resistance to precipitation by calcium and renders bile acids impermeable to cell membranes at neutral pH.[10] From the liver, bile acids are rapidly pumped into the biliary canaliculus by the bile-salt export pump. As bile flows through the bile duct to the gallbladder for storage, it is modified by the addition of a watery, bicarbonate-rich secretion from ductal epithelial cells.[11] This gives the bile solution a slightly alkaline pH. The conjugated bile acids are subsequently released from the gallbladder into the duodenum to aid in the digestion and absorption of fat, at a concentration of around 100 mM.[12,13] Anaerobic bacterial flora in the lower GI tract are responsible for the metabolism of bile, firstly through deconjugation, releasing free bile acids and secondly, they induce 7-α dehydroxylation to yield deoxycholic (DCA) and lithocholic acids (LCA) from cholic acid (CA) and chenodeoxycholic acid (CDCA), respectively (Figure 6.1).[14] Tertiary bile acids can result from further bacterial biotransformation, including ursodeoxycholic acid (UDCA) (Figure 6.1), which also enter the total bile acid pool.

Bile acids are "recycled" *via* the enterohepatic circulation, with less than 5% of the total bile acid pool entering the colon.[15,16] Bile acids are reabsorbed by ileum columnar epithelium cells and are transported back to the liver by the portal vein where they are extracted by hepatocytes.[13] Approximately 6–12 enterohepatic circulations occur daily.[17] Free bile acids, like DCA, are partly absorbed into the colon and enter the enterohepatic circulation, where they are

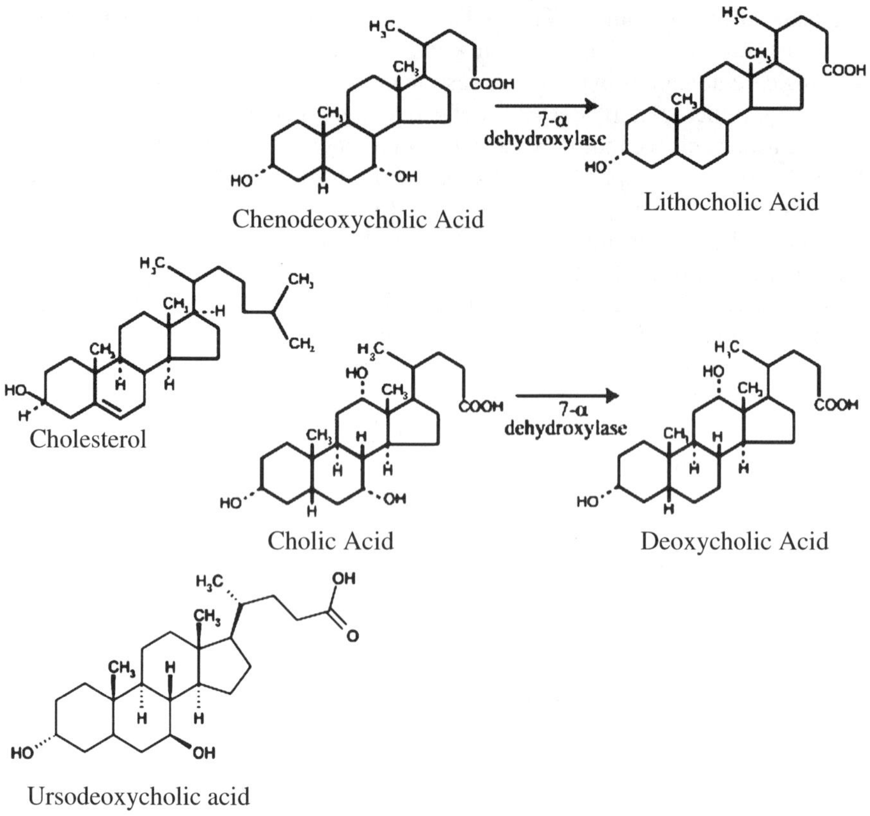

Figure 6.1 C_{24} bile acids are synthesised in the liver from cholesterol. The bacterial enzyme 7-α dehydroxylase converts CDCA into LCA and converts CA into DCA. UDCA a tertiary bile acid is also shown.

reconjugated in the liver and secreted in the bile.[17] Little of the LCA is re-absorbed as it is fairly insoluble.[14] Due to the efficient re-conjugation of bile acids from the enterohepatic circulation in the liver, there should, in theory, be minimal amounts of free (*i.e.* unconjugated) bile acids in the duodenum (and hence stomach and oesophagus).

Bile acids are frequently found in the stomach juice due to intestino-gastric reflux through the pyloric sphincter and their role in Barrett's oesophagus induction and OA development has been postulated for some time.[18] The alkaline bile is acidified by the stomach acid and the mixture of bile and acid is often refluxed further into the lower oesophagus during episodes of GORD. The presence of bile in the oesophagus is termed duodeno-gastro-oesophageal reflux (DGOR). The presence of bile acid in the stomach can also cause gastric mucosal damage, which can be a marker for increased risk of oesophageal damage and Barrett's oesophagus (and indeed OA) in GORD patients. Detailed work by Dixon *et al.*[19] has shown that a bile-reflux index (BRI), which measures bile induced damage in the stomach, is significantly associated with Barrett's metaplasia, thus providing evidence of a role for bile acids.

6.3   Bile Acids and Diet

It is important to understand that diet may play a major role in determining bile acid levels in individuals. In turn, bile acid levels may promote several forms of cancer including OA.

Firstly, bile acids are derivatives of cholesterol, therefore high cholesterol diets will lead to higher than normal levels of bile acids,[20] this has been estimated to translate into 2- to 5-fold increases in bile acid concentration in some individuals.[21] A link between diet and OA was previously identified following studies showing that OA induction by bile acid reflux in animal models was promoted by the supplementation of a high-fat diet.[22] Moreover, a recent paper has shown in animal models that high-fat diets not only increase bile acid production, but also alter the composition of the bile acids, leading to a promotion of OA development.[23] Therefore, it is important to take note of the impact of diet on GI tract cancers in general and in particular in relation to how diet can modulate the bile acid pool. This opens the door to dietary modulation of OA risk by reducing animal fat intake (reducing cholesterol) and hence overall bile acid reduction.

6.4   Duodeno-Gastro-Oesophageal Reflux (DGOR); Animal Models

Over the past 15 years there have been many animal studies that have proposed a link between DGOR and OA. This work has been invaluable in unpicking the relative contributions of acid and bile reflux to oesophageal damage and carcinogenesis. A lot of this work is summarised in a paper by Marshall *et al.*[24] and concludes that bile acid reflux generally worsens the oesophageal damage induced by acid reflux alone. Attwood and colleagues[25] showed that surgically enhanced duodenal reflux (through duodeno-oesophageal anastomosis) led to an increase in oesophageal cancer in rats over and above control rats and importantly, also over and above rats with enhanced gastro-oesophageal acid reflux (oesophago-gastroplasty). In this study 2 mutagenic agents (DMNM and MNAN) were used separately to promote oesophageal cancer. This study clearly showed duodenal reflux to be more carcinogenic than acid reflux alone. Ireland *et al.*,[26] using similar surgical techniques on rat models, showed elegantly that duodenal reflux promoted greater carcinogenesis induced by the mutagen MNAN. Intriguingly, rats with mixed acid and bile reflux had less tumour induction. The authors concluded that acid actually protects against the development of OA. A key study by Miwa *et al.*,[27] showed that rats with biliary diversions into the stomach, but not pancreatic diversions produced gastric cancer. This helped unpick the relative contributions of biliary and

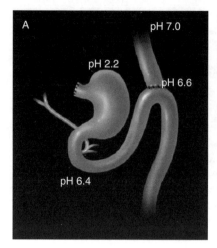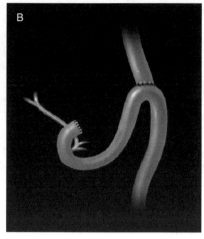

Figure 6.2   Two major surgical models of BO and OA, oesophagojejunostomy with gastric preservation (A) and oesophagojejunostomy with total gastrectomy (B). Figure 2A shows intra-luminal pH values in the animals one week after esophagojejunostomy and preservation of the stomach (adapted from Fein *et al.*[28]).

pancreatic agents from DGOR (duodeno-gastro-oesophageal reflux) and pointed the finger squarely at the biliary components. It is important to note that in several of these studies cancer was induced artificially with a mutagenic agent to increase tumourigenesis in these short-lived models. Nonetheless, these studies strongly implicate bile reflux in promoting oesophageal carcinogenesis.

A seminal rat surgical study in this regard was carried out by Fein *et al.*,[28] and is illustrated in Figure 6.2. This group showed that oesophagojejunostomy with gastric preservation inducing gastroduodenal reflux leads to BO and OA. Oesophagojejunostomy with total gastrectomy to induce oesophageal reflux of duodenal juice alone also results in BO and OA in this rat model without any carcinogen after 16 weeks. Importantly, 87% of animals with duodenal reflux induced by total gastrectomy and oesophagojejunostomy without carcinogen, developed columnar lining of the distal oesophagus.[28] Nearly half (48%) developed adenocarcinoma at the anastomotic site 16 weeks post-operatively. They concluded that duodenoesophageal reflux is carcinogenic in the rat model and that exogenous carcinogen is not necessary for cancer development in gastrectomised rats.[28]

Furthermore, Fein *et al.*[28] showed that at the anastomosis site in the animals with oesophagojejunostomy with gastric preservation, intra-luminal pH was 6.6, indicating that gastric acids are almost neutralised at the anastomosis site (Figure 6.2). Intraluminal pH values in the animals one week after oesophagojejunostomy and preservation of the stomach are shown in Figure 6.2.[28] Thus, it is possible that in the animal model using this surgical procedure gastric acid does not play a significant role and that bile acids are perhaps the more important factor in the development of BO/OA.[28]

6.5 Duodeno-Gastro-Oesophageal Reflux (DGOR); Clinical Data

The presence of bile acids in the oesophagus of patients with GORD has been known for some time.[29–31] In fact, in a recent study, over 90% of 172 endoscopy patients with either oesophagitis (48), Barrett's metaplasia (60) or neither (52) had bile in their gastric juice (and by inference in their refluxate).[32] Table 6.1 contains further findings from several studies, of DGOR in GORD patients. The refluxate studied in these GORD patients was often made up of mixtures of both stomach acid and bile (as well as other compounds such as stomach and pancreatic enzymes). In one study 13/30 (43%) GORD or Barrett's patients showed this type of mixed bile plus acid reflux,[33] whilst another stated that only 11% of 205 patients had mixed reflux.[31] However, in some cases the refluxate can be composed of bile without stomach acid. In Nehra's study, 6/30 (20%) patients showed bile reflux without acid,[33] in another study, this "bile only" reflux was noted in 16/32 patients.[34] Presumably "bile only" reflux into the oesophagus is a consequence of reduced gastric secretion either due to atrophy of the HCl producing gastric glands, or due to deliberate achlorhydria through acid-suppression medication as a treatment for GORD or dyspepsia.

DGOR is known to be more common in the supine position (*i.e.* at night-time)[29,30] and shows higher levels post-prandially,[30] due to the presence of food in the stomach triggering bile acid release from the gallbladder. The increase in DGOR in the supine position may mean that GORD patients have pools of refluxate in the lower oesophagus for extended periods of time during night-time. This night-time reflux may well contribute significantly to OA development.

Importantly, in terms of the development of OA, the actual bile acid concentration in the refluxate has also been linked to the presence of erosive oesophagitis and Barrett's metaplasia.[30,31,33,35] This suggests a causal relationship for bile, in the early stages of OA development. Studies have shown that DGOR worsens the severity of the oesophageal damage, with BO patients having the highest bile acid levels.[33,36–41] Not surprisingly, as the major treatment for GORD is acid suppression, it has been noted that in patients taking PPIs, the oesophageal refluxate is much more frequently nonacidic, with weakly acidic (90%) or weakly alkaline (10%) reflux being observed.[42] This can have important consequences on the bioactivity of any bile acids refluxed into the oesophagus (see later).

It is important to remember that bile acids are diverse molecules with very different activities and therefore pose different threats. The sub-sets of unconjugated bile acids (particularly secondary unconjugated bile acids) are most injurious. In our experience, the conjugated bile acids and the primary bile acid, cholic acid, are largely inactive, whereas the secondary bile acids DCA and LCA and the primary bile acid CDCA are potent inducers of DNA damage, toxicity and oncogenic signalling pathways. Of particular importance is the presence of the secondary bile acids (like DCA) in the oesophageal refluxate. It has been shown that not only does total bile acid concentration rise

Table 6.1 Evidence for bile reflux playing a role in Barrett's oesophagus and adenocarcinoma.

Reference	Methods	Results	Notes
Gotley *et al.*, 1988.[29]	$N = 55$, 45 GORD patients, 10 controls. HPLC for bile acid analysis.	Conjugated bile acids in 2/10 controls and 39/45 patients. Tending towards more bile in patients v controls. 11/39 had concentrations $> 200 \, \mu M$	Supine period had highest levels of bile acids. Nasopharyngeal intubation may increase saliva production and swallowing, hence diluting re-trieved refluxate. No unconjugated bile detected here.
Stein *et al.*, 1992.[31]	Alkaline reflux (pH > 7) recorded in the oesophagus as marker of DGOR. 17% of time pH > 7 as threshold. $N = 322$ (50 controls, 205 GORD patients, 67 non GORD symp-tomatic patients).	More alkaline reflux in the oesophagus of Barrett's patients or patients with oesophagitis or stricture, compared to controls (uncomplicated GORD, volun-teers or no GORD)	Limitations of using alkaline reflux, as saliva, food, dental infections, *etc*. can interfere with results. Defective sphincter pressure more common in Barrett's patients than GORD patients or controls. Of 205 GORD patients, 11% had both acid and bile reflux 89% had only acid reflux.
Iftikhar *et al.*, 1995.[35]	Aspiration study of bile from oesophagus, coupled to pH probe. HPLC for bile acid content. $N = 60$ patients, 30 with Barrett's oesophagus, 15 with oesophagitis and 15 controls.	More bile acid (total) in Barrett's and oesophagitis group than controls. 15/30 Barrett's patients had mM concentrations of total bile acid.	pH reading above 5 in the stomach similar in all 3 groups, therefore pH monitoring not associated with bile acid reflux, as this was sig-nificantly different between Bar-rett's patients and oesophagitis patients versus controls.
Kauer *et al.*, 1995.[98]	Aspiration study, oesophageal pH monitoring and bilirubin detection *via* optic fibre. OD Reading of 0.2 taken as threshold. $N = 25$ controls, 53 GORD patients.	22/53 GORD patients had acid only reflux. 31/53 had bile + acid reflux. Positive association between bile reflux and histology (*i.e.* metaplasia).	Majority of mixed reflux occurred at a pH of 4–7 (87%).
Kauer *et al.*, 1997.[30]	Oesophageal aspiration over 17 h. HPLC analysis of bile acids.	86% of GORD patients showed some bile in the oesophagus,	More bile aspirated post-prandially and in supine position.

Study	Methods	Results	Comments
	$N = 80$ (43 controls 19–48 years old, 37 GORD patients 21–67 years old).	compared to 58% of controls. More bile also in patients than controls. More bile acids in patients v controls in upright, post-prandial and supine periods.	Patients received refluxogenic liquid diet, that didn't interfere with bilitec device.
Vaezi and Richter, 1997.[34]	Bilitec 2000 for 18–24 h. > 0.14 bilirubin values set as cut-off. Partial gastrectomy patients suffering with GORD. $N = 32$ patients, 20 controls from prior study	9 patients had both acid and bile reflux, 16 patients had bile reflux alone and 7 had neither acid nor bile reflux. Therefore 25/32 (78%) of these patients had bile reflux, whereas 9/32 (28%) had acid reflux. There was a significant association between mixed reflux and oesophageal injury.	6/30 patients had bile reflux without acid. 13/30 had mixed reflux.
Nehra et al., 1999.[33]	18-h sampling with a peristaltic pump for aspiration of oesophageal samples, coupled to pH probe. HPLC analysis of aspirated samples. $N = 40$	Increasing bile acid profile in Barrett's/stricture patients. Controls had negligible bile acid reflux. Significant increase in unconjugated bile acids in Barrett's and erosive groups.	
Mainie et al., 2006.[42]	Multi-channel intra-luminal impedance, coupled to pH monitoring. $N = 168$ patients.	69 symptomatic patients, 16 (11%) had acidic reflux and 53 (37%) has nonacid reflux. Non-acid reflux was usually weakly acidic (90% of the time) or weakly alkaline (10% of the time).	Patients were taking twice daily PPI.
Fein et al., 2006.[115]	Bilitec 2000. $N = 341$. $OD_{453} > 0.25$ applied as a threshold. 189 patients with nonerosive oesophagitis, 76 with Barrett's oesophagus and 76 with erosive oesophagitis, 35–41 healthy controls.	More bile present in both stomach and oesophagus of GORD patients versus controls. Highest levels of bile abundance in Barrett's patients (but not significant). Of the 84 patients with increased bile in the oesophagus and stomach, 89% had erosive oesophagitis or Barrett's.	Bile hardly ever seen in oesophagus of healthy controls. Of the 173 patients with bile in the oesophagus, ~50% had no detectable bile in the stomach (~50% did). Therefore, bile in oesophagus, but not stomach in 50% of cases.

Table 6.1 (*Continued*).

Reference	Methods	Results	Notes
Darragh *et al.*, 2006.[32]	Gas chromatography on gastric aspirates, $N = 172$	Bile present in 92% (158/172) of patient samples. No difference in bile content between patients with Barrett's (60), oesophagitis (48) or controls ($N = 52$).	12 patients without clinical information. All patients underwent endoscopy for GORD symptoms.
Wolfgarten *et al.*, 2007.[44]	Bilitec 2000, measuring bilirubin at OD 453 nm. 0.25 reading as cutoff for noxious bile levels. $N = 64$	More bile in stomach of GORD patients than controls. Percentage time bile in oesophagus 7.8% for Barrett's patients, 3.5% for oesophagitis patients, 0% for controls.	First report using genuine age-matched controls. White diet to reduce noise from food matter. Also, more (and longer) bile reflux into stomach during supine (sleeping) position. Acid reflux also greater in patients v controls.
Nason *et al.*, 2007.[45]	Measured bile in gastric aspirates as surrogate for DGOR. HPLC used to identify bile acids $N = 72$ GORD cases with metaplasia and 71 controls with GORD only.	No significant differences in bile-acid content between cases and GORD controls. No correlation with presence of metaplasia or oesophagitis and bile acid content.	Limited bile content could be due to fasting status of the patients. No control group.

in Barrett's patients and GORD patients relative to controls,[30,33,35] but the levels of the more injurious bile acids like DCA also rises.[33] The identification of the candidate bile acid sub-types most closely linked to oesophageal carcinogenesis has been of the utmost importance in terms of understanding the mechanisms involved and in the search for therapies and screening tools to assess individual cancer risks.

6.6 Measurement of Bile in the Oesophagus

Some early studies of bile acid reflux in GORD patients measured alkalinity in the oesophagus as a measure of DGOR.[31] Whilst this gave some useful information, it had some basic flaws due to interference from saliva, food particulates, *etc.* Indeed, in a study using HPLC to measure bile acid reflux, Iftikhar *et al.*,[35] showed that bile content increased in Barrett's patients relative to controls, whilst pH did not differ. Hence, more specific tests were sought. Alkalinity measurements were superseded by devices such as the Bilitec device that measures bilirubin absorption at 453 nm. As bilirubin is excreted from the liver in bile, it acts as a useful marker of bile content.[34,43,44] The Bilitec device has widely been used to study bile acid content in large cohorts of patients and has shown increased bile reflux in Barrett's oesophagus patients versus controls.

The most exact measurements of DGOR however, have used direct assessment of bile acid concentration in aspirates by HPLC and gas chromatography (GC).[29,30,32,33,35] Other enzymatic assays for bile acid measurement in aspirates have suffered from some cross-reactivity to other sterol groups. The HPLC/GC methods have been employed to assess the relative concentrations of each of the bile acids in groups of patients. Table 6.2 summarises some of the available data on individual bile acid concentrations in refluxate. This kind of data has been invaluable in assigning roles to individual bile acids in the pathogenesis of Barrett's metaplasia and OA. Obtaining oesophageal refluxates for HPLC/GC analysis is however, time consuming (up to 24 h collection required), impractical and imposes severely on the patients. Indeed, it is questionable whether many research ethics committees would now allow such invasive studies, and indeed whether patients would willingly volunteer for them. Therefore, some researchers have recently used the collection of gastric juice (during normal endoscopy) as a surrogate for the oesophagus with the assumption that any bile acids present in the stomach of a refluxing patient will undoubtedly enter the oesophagus.[32,45] This is a very pragmatic way forward to study the role of bile acids in OA development.

These above-mentioned studies quantifying bile acid reflux have been fundamental to allowing *in vitro* analysis of bile acid effects at physiological doses (Table 6.2). These *in vitro* studies have crucially, identified molecular mechanisms important in bile-driven carcinogenesis. These molecular events will undoubtedly be important in future years as drug-able targets and as biomarkers of cancer risk. Prior to bile acid quantification in the refluxate, there

Table 6.2 Amount of bile acids refluxed into the oesophagus.

Reference	Methods	Average amount of bile acids		Peak amounts of bile acids
Stein *et al.*, 1994.[116]	Aspiration and enzymatic assay for bile acids.	Controls 5 µM 12 µM 7 µM	GORD (total bile) 12 µM (upright) 72 µM (post-prandial) 107 µM (supine)	
Kauer *et al.*, 1997.[30]	HPLC for bile acids. 43 healthy volunteers, 37 GORD patients. Oesophageal aspiration over 17 h.	GCA 98 µM TCA 13 µM GDCA 30 µM TDCA 5 µM GCDCA 69 µM TCDCA 13 µM GLCA1.3 µM		
Nehra *et al.*, 1999.[33]	HPLC analysis of oesophageal aspirates (15 h). $N = 40$. 10 volunteers, 10 GORD with minimal injury, 10 with erosive oesophagitis, 10 with Barrett's oesophagus/ stricture.	CA 34 and 25 µM[a] TCA 21 and 39 µM[a] GCA 21 and 27 µM[a] DCA 2 and 0 µM[a] TDCA 0 and 5.5 µM[a] GDCA 2.5 and 0 µM[a] TLCA 0 and 6 µM[a]		150 and 126 µM[a] 198 and 131 µM[a] 88 and 95 µM[a] 230 and 282 µM[a] 157 and 410 µM[a] 285 and 72 µM[a] 0 and 482 µM[a]
Darragh *et al.*, 2006.[32]	Gas chromatography on gastric aspirates, $N = 172$	Conj CA 118 µM CA 5 µM Conj CDCA 112 µM CDCA 3 µM Conj DCA 63 µM DCA 3 µM Conj LCA 17 µM LCA 2 µM Conj UDCA 13 µM UDCA 18 µM		2447 µM 211 µM 3655 µM 121 µM 1592 µM 115 µM 515 µM 82 µM 860 µM 978 µM

[a]All figures show erosive oesophagitis and Barrett's metaplasia readings, respectively.

was little idea of the physiological dose range and hence some early *in vitro* studies on bile acids effects used superphysiological doses (mM ranges) which clearly do not reflect the situation *in vivo*. Knowing the physiological doses of bile acids commonly seen in oesophageal refluxates, allows mimicking of the effects of bile acids *in vitro*.

6.7 Mechanisms behind the Carcinogenicity of Bile Acids

Recently, there has been increasing evidence pointing towards bile acids being carcinogens in their own right. As well as the accumulating clinical evidence, there is good experimental data indicating that they are capable of inducing DNA damage[46,47] and activating oncogenic signalling pathways.[48–50] It is likely that bile acids cause these effects, through their ability to induce intra-cellular reactive oxygen species (ROS).[47,51–53] Bile acid induction of oxidative stress is thought to be linked to their membrane-damaging role (as detergents). Mitochondrial membrane damage may well lead directly to increased ROS leakage into the cytoplasm.[54] Oxidative stress can result in induced mutagenesis in human cells,[47,55,56] which could lead to mutations, resulting in the aberrant expression of oncogenes and tumour-suppressor genes, and eventually cancer. Therefore, the investigation of the role of ROS in the pathogenesis of BO and OA, and the potential role of anti-oxidants in blocking these processes are fertile areas of research.

As early as 1940 the bile acid DCA was proposed to be carcinogenic in mice.[57] More recently, administration of CDCA to mice with a germline mutation in Apc (Min/+) resulted in an increase in duodenal tumours.[58] In rat experiments, bile acids such as LCA and DCA were shown to increase tumourigenesis by known carcinogens, including MNNG. However, the bile acids *per se* were not always found to be carcinogens in these studies, failing to induce colon tumours themselves.[20,59,60] This may be explained by the lack of time available to induce tumours in these short-lived animal models.

In the oesophagus, a key molecular event involved in neoplastic development is activation of the transcription factor nuclear factor kappa B (NF-κB). In a study looking at the activation of NF-κB in Barrett's epithelium and adenocarcinoma, it was found that NF-κB was increasingly activated in patients during histological progression (0% of patients with squamous epithelium had active NF-κB, 40% of patients with Barrett's epithelium and 61% of resected tumours displayed NF-κB activity). Furthermore NF-κB activity was linked to later-stage cancers and was inversely correlated with major or complete pathologic responses to neoadjuvant chemotherapy and radiation therapy.[61] NF-κB activity is linked to many types of cancer[62] and this is thought to be due to its ability to switch on an array of genes, whose protein products promote carcinogenesis (*e.g.* BCL-XL, XIAP, COX-2, IL-8). Our group have found that immunohistochemistry (IHC) was effective at monitoring overall NF-κB activity and IL-8 abundance in Barrett's metaplasia and OA. Significant increases in nuclear NF-κB activity and IL-8 abundance were observed across the histological series by IHC analysis.[63]

A major mechanistic link between bile acids and OA is apparent from studies showing that bile acids (DCA in particular) are capable of activating NF-κB and causing up-regulation of NF-κB controlled genes *in vitro*, mimicking the observed NF-κB activation *in vivo*.[61,64] Results from our lab have shown that physiological levels of DCA (*i.e.* 100–300 µM) were capable of activating NF-κB in OE33 cells resulting in NF-κB target gene expression.[50] NF-κB activation by DCA occurs at neutral pH, but less so at an acidic pH in the OE33 oesophageal cell line.[50] Hence, bile (DCA) and to a lesser extent acid, may be responsible for the activation of NF-κB in the oesophagus of refluxing patients.

Importantly many of the molecular effects of bile acids can be blocked with anti-oxidants, supporting a role for ROS as key intermediaries. For example, we have shown that DCA (but not conjugated bile acids or cholic acid) can induce chromosome damage *in vitro*. However, this chromosome damage can be effectively blocked by the presence of equimolar amounts of the anti-oxidant vitamin C.[47] Furthermore, DCA's induction of NF-κB in OE33 cells is effectively blocked by various anti-oxidants like vitamin C, curcumin, *etc.* (unpublished results). Hence, this raises the possibility that anti-oxidant supplementation can block many of the deleterious consequences of bile acid exposure *in vivo* and this may slow the development of cancer in these patients. In fact, it has been well reported that patients with Barrett's oesophagus who have a diet rich in anti-oxidants have a 40–50% reduced risk of developing OA.[65] Hence dietary supplementation with anti-oxidants may be a promising treatment for patients with GORD or Barrett's oesophagus.

6.8   The Interplay between Acid and Bile in GORD

Acid is undoubtedly important in the pathogenesis of GORD and given the ability of acid stimuli to activate oncogenic signalling,[66–68] it is not possible to rule out acid as an oesophageal carcinogen in its own right. However, growing evidence indicates that bile acids are at least equally important in OA development. Moreover, due to the fact that acid suppression hasn't slowed OA incidence, one could argue that bile acids may be superseding acidity as the causative factor in OA. If bile and not gastric acid is the most important carcinogenic insult to the oesophagus in GORD patients, what is the nature of the interaction between acid and bile, and what is the consequence of removing stomach acidity through medical intervention in GORD patients?

6.9   Deconjugation of Bile Acids in Nonacidic Stomach

The generation of unconjugated bile acids in an alkaline milieu in the stomach is probably the most important mechanism promoting bile-driven carcinogenesis in acid-suppressed patients. The presence of unconjugated bile acids at near-neutral pH in the refluxate will exacerbate oesophageal mucosal damage.[69,70] As mentioned earlier, due to an efficient conjugation step in the liver, there should be no free bile acids leaving the gallbladder for the duodenum. However, acid suppression, resulting in a higher pH in the stomach and immediate environment, may be more favourable to bacterial growth and subsequent deconjugation of any released bile acids. These unconjugated bile acids are more injurious to the mucosa of both the stomach and the oesophagus. Several studies have shown that acid-suppression therapy allows gastric and small intestinal bacterial overgrowth[71–76] some groups having shown that this allows bacterial deconjugation to actually occur.[77,78] As is shown in Table 6.2, unconjugated bile acids are common in the stomach and the oesophagus, hence supporting a role for de-conjugation of bile acids in the stomach.

6.10 Effect of Acidity on Bile Acid Solubility

The ability of bile acids to damage the oesophageal mucosa depends, not only on the conjugation state of bile acids, but also on the pH of the refluxate. The solubility of bile acids and their toxic effects are known to vary with pH, depending on the degree of ionisation. Bile acids are most injurious to cells when they are un-ionised and soluble. The pKa of all the common natural C24 unconjugated bile acids is close to 5.1, whereas glycine- and taurine-conjugated bile acids have a pKa of around 4 and 2, respectively.[79] Harmon *et al.*,[80] showed in rabbit tissue that unconjugated DCA and CA caused oesophageal damage at neutral pH, whereas their taurine-conjugated counterparts caused damage at pH 2. Unconjugated bile acids and glycine conjugates whose pKa values are more than 6 and 4, respectively, precipitate in the acidic stomach, whereas taurine conjugates are freely soluble even at pH 2.[10] Hence, due to the fact that unconjugated bile acids are thought to be the most injurious, it appears that acid in the stomach is a natural protective mechanism against these agents. It has been suggested for some time that acid suppression will keep the damaging bile acids in solution and hence promote their activity.[30] Acid-suppression medication will remove the protective effect of acidity and will promote the damage done by free bile acids like DCA. This is somewhat supported by data showing that acid suppression increases the mutagenicity of gastric juice.[81] A recent report has also highlighted the possibility that acid itself can suppress proliferation in oesophageal cells,[82] thus supporting a protective role for stomach acid.

In our experience, at neutral pH, conjugated bile acids have less or no biological effects when compared with unconjugated dihydroxy bile acids, which are more bioactive. It is possible that conjugated bile acids in their un-ionised and soluble state at low pH, act in synergism with acid causing mucosal damage. However, if acid-suppression therapy results in a more neutral intra-gastric pH milieu, then un-ionised and soluble unconjugated bile acids could permeate the cell membrane and accumulate within the cell causing mucosal damage.[83] Animal models of oesophagitis have shown that exposure to low lumenal bile acid concentrations can cause increased mucosal permeability to a variety of ions and molecules without causing gross morphologic damage. One study found that the bile acids CDCA and TCA can accumulate in rabbit oesophageal mucosa to levels as high as seven times the initial lumenal concentration.[84] These findings are consistent with a combination of intra-cellular entrapment of the bile acids due to intra-cellular ionisation and bile acid binding to cellular membrane components being the mechanisms by which bile acids accumulate in cells. Bile acid accumulation may explain how relatively low lumenal concentrations of bile acid can be damaging to the gastro-intestinal mucosa.[85]

6.11 Acid-Suppression Therapy and Risks of OA

PPIs are currently in wide use for the management of GORD, have a good efficacy and are presumed safe. However, there are contrasting reports on the effect of complete acid suppression on DGOR. Some studies of Barrett's patients have reported that complete acid suppression prevented exposure of the oesophagus to duodenal contents,[37,40,86] whereas other studies propose that there is persisting bile reflux despite pharmacological intervention.[87–89] A recent study reported that around 80% of BO patients on PPIs showed a normal oesophageal pH profile. However, 60% showed abnormal oesophageal exposure to bile in their oesophagus.[90]

An alternative strategy to treat GORD and to reduce neoplastic risk in Barrett's patients is to surgically tighten the gastro-oesophaegeal junction using a procedure like a Nissen fundoplication. Obviously, in practice this is not a feasible treatment for all GORD patients, but could be used for high-risk progressors. In theory this should reduce the oesophageal exposure to both acid and bile (and other constituents of refluxate). Therefore, one would expect that surgical treatment of GORD/Barrett's would be superior to medical treatment in terms of preventing OA development. However, there is little evidence to support this.[91] This is perhaps due to short follow-up times, small numbers of patients in the studies, or the surgical procedure being nonconstant across studies. Another important confounding factor here is that the surgical groups tend to include the patients with the worst symptoms of reflux; hence there is poor matching in these studies for patient symptoms. It has also been shown recently that some patients still suffer with GORD post-surgery,[92] thus confounding the comparison with medication further.

Supporting evidence for a role for bile reflux in OA comes from other studies of a surgical nature. There is a reported increase in OA development in patients after cholecystectomy.[93] This operation can, in theory, lead to increased bile-acid levels in the duodenum (and hence stomach and oesophagus). The correlation between OA and cholecystectomy supports the concept that DGOR is important in OA development.

There is no consistent clinical evidence to support a role for acid suppression in the prevention of BO patients progressing to adenocarcinoma. Acid suppression has been reported to result in the generation of squamous islands, although there is no significant evidence of a reduction in the overall length of the BO.[94] There is also a reported reduction in the development of dysplasia in patients on acid suppression.[95] However, several studies have concluded that acid-suppression therapy may actually have a deleterious effect[96–98] and the association between the increasing incidence of OA and the introduction of acid-suppression therapy has been postulated.[99] Indeed, epidemiological studies have shown a dramatic increase in rates of OA since H$_2$RAs and PPIs were first introduced. However, it is possible that this is coincidental. Lagergren *et al.*[100] found that the use of acid-suppression therapy was greater in those who developed OA and was an independent risk factor, leading to a 3-fold increased risk. In a recent nested case control study in the UK, the same group found that gastric-acid suppression is a marker of increased risk of OA (3-fold increased risk for patients on acid suppression, leading to 5-fold increased risk in long-term users). However, they concluded that these associations were likely explained by the underlying treatment indication being a risk factor for the cancer rather than an independent harmful effect of these agents.[101] Other studies have found an increase in oesophageal cancer rates in patients using H_2 receptor antagonists, although there was no differentiation between squamous cancer and adenocarcinoma.[102,103] Conversely, another study found no increased risk of oesophageal cancer.[104] The UK's current AspECT trial, which contains a low-dose versus high-dose PPI arm (as well as ± aspirin), may ultimately produce some answers to the questions surrounding acid suppression and oesophageal cancer risk. It is unfortunate that there was no possibility of including a "no PPI" arm in this trial.

Acid-suppression therapy may also possibly contribute to oesophageal carcinogenesis through 3 additional mechanisms. Firstly, acid suppression leads to hyper-gastrinemia, an elevation in the levels of the gastrin hormone. Gastrin is known to induce oesophageal proliferation, reduce apoptosis[105] and up-regulate Cox-2.[106] Secondly, the bile acid concentration may be increased in the ensuing refluxate after acid suppression, due to a reduction in the overall gastric juice volume. Acid suppression is known to reduce overall duodenal reflux values (as measured by Bilitec, for example) as well as suppressing acid reflux in the oesophagus.[86] However, in the smaller volume of refluxate actually reaching the oesophagus, the local concentrations of bile acids may be higher and hence more damaging. Finally, acid suppression has also been shown to reduce the bioavailablity of dietary vitamin C.[107] Given that ROS are implicated as intermediates in bile acid activity, the loss of anti-oxidant capacity may increase levels of bile-driven DNA damage and oncogenic signalling.

6.12 Bile Acid Therapy

Not all bile acids are potentially harmful; indeed some bile acids can antagonise the negative effects of injurious bile acids. For example, UDCA, which is currently in use in the treatment of cholestatic liver disease[108–111] and is thought to have chemoprotective properties in colon cancer.[51,112] A recent study has shown that the bile acids DCA and UDCA have distinct effects on the DNA binding of NF-κB in colon-cancer cell lines. UDCA was found to inhibit IL-1β induced NF-κB binding, by blocking IκBα degradation induced by DCA.[113] In another study UDCA was also found to inhibit IκBα degradation, blocking the expression of NF-κB-dependent genes in microglia cells when activated by beta-amyloid peptide.[114] Hence, UDCA may also provide the possibility of a preventative treatment for patients with Barrett's oesophagus.

6.13 Conclusions

There is a burgeoning weight of evidence emerging (some of which has been around for some time) suggesting that bile acids are potent oesophageal carcinogens. This data comes from animal models of oesophageal carcino-genesis, from clinical studies linking bile acid content in the oesophagus to histological changes, as well as from mechanistic cell-biology studies high-lighting the ability of bile acids to activate oncogenic signalling and induce DNA damage. This may well tip the balance in favour of bile acids in the ar-gument as to which reflux constituent is most important in oesophageal cancer. Acid reflux may, in some patients, solely be responsible for oesophageal damage and histological progression. However, in most cases, it can be argued, concurrent bile reflux content worsens the situation and this is important, given that bile is present in most GORD patient's reflux, ranging from 51% of pa-tients,[115] to 78%,[34] to 86%,[29,30] to 92%.[32]

As pointed out by Dixon and colleagues;[19] "if a direct role for refluxed bile derivatives in carcinogenesis becomes accepted, therapy aimed principally at acid reduction, cannot be expected to eliminate cancer risk in Barrett's oesophagus." However, not only is acid suppression likely to be ineffective in preventing bile acid-driven carcinogenesis, it could also actually promote cancer development by providing a less acidic environment favouring bile acid activity.

Hence, in summary;

- Bile acids are centrally involved as carcinogens in OA. The levels of bile acids inducing OA may be dietary in origin (high levels of animal-fat intake).
- Screening for bile acid (by sub-type and concentration) is warranted as a biomarker for OA development.
- Acid suppression is probably responsible for deconjugation of bile acids in the upper GI tract and hence oesophageal exposure to free bile acids.
- Acid suppression also promotes the activity of the most damaging sub-types of bile acids; stomach acid is a natural defence against this.
- Reactive oxygen species play a central role in bile acid bioactivity. Hence anti-oxidant supplementation can correct some of the deleterious effects of acid suppression.
- Other forms of possible treatment include bile acid sequestrants and UDCA.

Acknowledgements

We are extremely grateful to Katerina Dvorak for critically reading this document and for providing Figure 6.2 presented here.

References

1. M. Pera, C. Manterola, O. Vidal and L. Grande, *J. Surg. Oncol.*, 2005, **92**, 151–159.
2. G. R. Locke III, N. J. Talley, S. L. Fett, A. R. Zinsmeister and L. J. Melton III, *Gastroenterology*, 1997, **112**, 1448–1456.

3. J. Ronkainen, P. Aro, T. Storskrubb, S. E. Johansson, T. Lind, E. Bolling-Sternevald, M. Vieth, M. Stolte, N. J. Talley and L. Agreus, *Gastroenterology*, 2005, **129**, 1825–1831.
4. J. A. Jankowski, N. A. Wright, S. J. Meltzer, G. Triadafilopoulos, K. Geboes, A. G. Casson, D. Kerr and L. S. Young, *Am. J. Pathol.*, 1999, **154**, 965–973.
5. A. T. Prach, T. A. MacDonald, D. A. Hopwood and D. A. Johnston, *Lancet*, 1997, **350**, 933.
6. H. Pohl and H. G. Welch, *J. Natl. Cancer Inst.*, 2005, **97**, 142–146.
7. G. K. Reeves, K. Pirie, V. Beral, J. Green, E. Spencer and D. Bull, *B.M.J.*, 2007. [Epub ahead of print].
8. C. N. Falany, M. R. Johnson, S. Barnes and R. B. Diasio, *J. Biol. Chem.*, 1994, **269**, 19375–19379.
9. M. Norlin and K. Wikvall, *Curr. Mol. Med.*, 2007, **7**, 199–218.
10. A. F. Hofmann and K. J. Mysels, *J. Lipid Res.*, 1992, **33**, 617–626.
11. M. H. Nathanson and J. L. Boyer, *Hepatology*, 1991, **14**, 551–566.
12. S. Perwaiz, B. Tuchweber, D. Mignault, T. Gilat and I. M. Yousef, *J. Lipid Res*, 2001, **42**, 114–119.
13. G. A. Kullak-Ublick, B. Stieger and P. J. Meier, *Gastroenterology*, 2004, **126**, 322–342.
14. M. J. Hill, *Mutat. Res.*, 1990, **238**, 313–320.
15. R. N. Redinger, *Can. J. Gastroenterol.*, 2003, **17**, 265–271.
16. M. Muller and P. L. Jansen, *Am. J. Physiol.*, 1997, **272**, G1285–1303.
17. F. M. Nagengast, M. J. Grubben and I. P. van Munster, *Eur. J. Cancer*, 1995, **31A**, 1067–1070.
18. C. G. Bremner and R. J. Mason, *Brit. J. Surg.*, 1993, **80**, 1374–1376.
19. M. F. Dixon, P. M. Neville, N. P. Mapstone, P. Moayyedi and A. T. Axon, *Gut*, 2001, **49**, 359–363.
20. B. S. Reddy, K. Watanabe, J. H. Weisburger and E. L. Wynder, *Cancer Res.*, 1977, **37**, 3238–3242.
21. N. Breuer and H. Goebell, *Klin. Wochenschr*, 1985, **63**, 97–105.
22. G. W. Clark, T. C. Smyrk, S. S. Mirvish, M. Anselmino, Y. Yamashita, R. A. Hinder, T. R. DeMeester and D. F. Birt, *Ann. Surg. Oncol.*, 1994, **1**, 252–261.
23. K. H. Chen, K. Mukaisho, H. Sugihara, Y. Araki, G. Yamamoto and T. Hattori, *Cancer Sci.*, 2007, **98**, 1683–1688.
24. R. E. Marshall, A. Anggiansah and W. J. Owen, *Brit. J. Surg.*, 1997, **84**, 21–28.
25. S. E. Attwood, T. C. Smyrk, T. R. DeMeester, S. S. Mirvish, H. J. Stein and R. A. Hinder, *Surgery*, 1992, **111**, 503–510.
26. A. P. Ireland, J. H. Peters, T. C. Smyrk, T. R. DeMeester, G. W. Clark, S. S. Mirvish and T. E. Adrian, *Ann. Surg.*, 1996, **224**, 358–370; discussion 370–351.
27. K. Miwa, H. Hasegawa, T. Fujimura, H. Matsumoto, R. Miyata, T. Kosaka, I. Miyazaki and T. Hattori, *Carcinogenesis*, 1992, **13**, 2313–2316.

28. M. Fein, J. H. Peters, P. Chandrasoma, A. P. Ireland, S. Oberg, M. P. Ritter, C. G. Bremner, J. A. Hagen and T. R. DeMeester, *J. Gastrointest. Surg.*, 1998, **2**, 260–268.

29. D. C. Gotley, A. P. Morgan and M. J. Cooper, *Brit. J. Surg.*, 1988, **75**, 587–590.

30. W. K. Kauer, J. H. Peters, T. R. DeMeester, H. Feussner, A. P. Ireland, H. J. Stein and R. J. Siewert, *Surgery*, 1997, **122**, 874–881.

31. H. J. Stein, A. P. Barlow, T. R. DeMeester and R. A. Hinder, *Ann. Surg.*, 1992, **216**, 35–43.

32. J. Darragh, M. Hunter, E. Pohler, L. Nelson, J. F. Dillon, R. Nenutil, B. Vojtesek, P. E. Ross, N. Kernohan and T. R. Hupp, *Febs. J.*, 2006, **273**, 1930–1947.

33. D. Nehra, P. Howell, C. P. Williams, J. K. Pye and J. Beynon, *Gut*, 1999, **44**, 598–602.

34. M. F. Vaezi and J. E. Richter, *Gut*, 1997, **41**, 297–302.

35. S. Y. Iftikhar, S. Ledingham, D. F. Evans, S. W. Yusuf, R. J. Steele, M. Atkinson and J. D. Hardcastle, *Gut*, 1995, **37**, 465–470.

36. C. Iascone, T. R. DeMeester, A. G. Little and D. B. Skinner, *Arch. Surg.*, 1983, **118**, 543–549.

37. G. Champion, J. E. Richter, M. F. Vaezi, S. Singh and R. Alexander, *Gastroenterology*, 1994, **107**, 747–754.

38. M. T. Caldwell, P. Lawlor, P. J. Byrne, T. N. Walsh and T. P. Hennessy, *Brit. J. Surg.*, 1995, **82**, 657–660.

39. W. K. Kauer, P. Burdiles, A. P. Ireland, G. W. Clark, J. H. Peters, C. G. Bremner and T. R. DeMeester, *Am. J. Surg.*, 1995, **169**, 98–103; discussion 103–104.

40. M. F. Vaezi and J. E. Richter, *Surgery*, 1995, **117**, 699–704.

41. D. Nehra, P. Howell, J. K. Pye and J. Beynon, *Brit. J. Surg.*, 1998, **85**, 134–137.

42. I. Mainie, R. Tutuian, A. Agrawal, D. Adams and D. O. Castell, *Brit. J. Surg.*, 2006, **93**, 1483–1487.

43. M. Fein, K. H. Fuchs, T. Bohrer, S. M. Freys and A. Thiede, *Dig. Dis. Sci.*, 1996, **41**, 216–225.

44. E. Wolfgarten, B. Putz, A. H. Holscher and E. Bollschweiler, *J. Gastrointest. Surg.*, 2007, **11**, 479–486.

45. K. S. Nason, D. C. Farrow, G. Haigh, S. P. Lee, M. P. Bronner, S. N. Rosen and T. L. Vaughan, *Interact. Cardiovasc. Thorac. Surg.*, 2007, **6**, 304–307.

46. A. J. Jolly, C. P. Wild and L. J. Hardie, *Mutagenesis*, 2004, **19**, 319–324.

47. G. J. Jenkins, F. R. D'Souza, S. H. Suzen, Z. S. Eltahir, S. A. James, J. M. Parry, P. A. Griffiths and J. N. Baxter, *Carcinogenesis*, 2007, **28**, 136–142.

48. F. Zhang, K. Subbaramaiah, N. Altorki and A. J. Dannenberg, *J. Biol. Chem.*, 1998, **273**, 2424–2428.

49. C. Tselepis, I. Perry, C. Dawson, R. Hardy, S. J. Darnton, C. McConkey, R. C. Stuart, N. Wright, R. Harrison and J. A. Jankowski, *Oncogene*, 2002, **21**, 6071–6081.

50. G. J. Jenkins, K. Harries, S. H. Doak, A. Wilmes, A. P. Griffiths, J. N. Baxter and J. M. Parry, *Carcinogenesis*, 2004, **25**, 317–323.
51. M. Venturi, R. J. Hambly, B. Glinghammar, J. J. Rafter and I. R. Rowland, *Carcinogenesis*, 1997, **18**, 2353–2359.
52. Y. Fang, S. I. Han, C. Mitchell, S. Gupta, E. Studer, S. Grant, P. B. Hylemon and P. Dent, *Hepatology*, 2004, **40**, 961–971.
53. H. Bernstein, H. Holubec, C. Bernstein, N. A. Ignatenko, E. Gerner, K. Dvorak, D. Besselsen, K. A. Blohm-Mangone, J. Padilla-Torres, B. Dvorakova, H. Garewal and C. M. Payne, *Dig. Dis. Sci.*, 2007, **52**, 628–642.
54. K. Dvorakova, C. M. Payne, L. Ramsey, H. Bernstein, H. Holubec, M. Chavarria, C. Bernstein, R. E. Sampliner, C. Riley, A. Prasad and H. Garewal, *Am. J. Gastroenterol.*, 2005, **100**, 424–431.
55. B. Glinghammar, H. Inoue and J. J. Rafter, *Carcinogenesis*, 2002, **23**, 839–845.
56. T. G. Rossman and E. I. Goncharova, *Mutat. Res.*, 1998, **402**, 103–110.
57. J. W. Cook, E. L. Kennaway and N. M. Kennaway, *Nature*, 1940, **145**.
58. N. N. Mahmoud, A. J. Dannenberg, R. T. Bilinski, J. R. Mestre, A. Chadburn, M. Churchill, C. Martucci and M. M. Bertagnolli, *Carcinogenesis*, 1999, **20**, 299–303.
59. T. Narisawa, N. E. Magadia, J. H. Weisburger and E. L. Wynder, *J. Natl. Cancer Inst.*, 1974, **53**, 1093–1097.
60. B. S. Reddy, T. Narasawa, J. H. Weisburger and E. L. Wynder, *J. Natl. Cancer Inst.*, 1976, **56**, 441–442.
61. M. M. Abdel-Latif, J. O'Riordan, H. J. Windle, E. Carton, N. Ravi, D. Kelleher and J. V. Reynolds, *Ann. Surg.*, 2004, **239**, 491–500.
62. B. Rayet and C. Gelinas, *Oncogene*, 1999, **18**, 6938–6947.
63. G. J. Jenkins, J. Mikhail, A. Alhamdani, T. Brown, S. Caplin, J. Manson, R. Bowden, N. Toffazal, P. Griffiths, J. Parry and J. Baxter, *J. Clin. Pathol.*, 2007, **60**, 1232–1237.
64. J. M. O'Riordan, M. M. Abdel-latif, N. Ravi, D. McNamara, P. J. Byrne, G. S. McDonald, P. W. Keeling, D. Kelleher and J. V. Reynolds, *Am. J. Gastroenterol.*, 2005, **100**, 1257–1264.
65. P. Terry, J. Lagergren, W. Ye, O. Nyren and A. Wolk, Antioxidants and cancers of the esophagus and gastric cardia, *Int. J. Cancer.*, 2000, **87**, 750–754.
66. R. F. Souza, K. Shewmake, L. S. Terada and S. J. Spechler, *Gastroenterology*, 2002, **122**, 299–307.
67. C. Morgan, W. Alazawi, P. Sirieix, T. Freeman, N. Coleman and R. Fitzgerald, *Am. J. Gastroenterol.*, 2004, **99**, 218–224.
68. K. Dvorak, C. M. Payne, M. Chavarria, L. Ramsey, B. Dvorakova, H. Bernstein, H. Holubec, R. E. Sampliner, N. Guy, A. Condon, C. Bernstein, S. B. Green, A. Prasad and H. S. Garewal, *Gut*, 2007, **56**, 763–771.
69. R. E. Marshall, A. Anggiansah, W. A. Owen and W. J. Owen, *Gut*, 1997, **40**, 182–187.

70. J. Theisen, D. Nehra, D. Citron, J. Johansson, J. A. Hagen, P. F. Crookes, S. R. DeMeester, C. G. Bremner, T. R. DeMeester and J. H. Peters, *J. Gastrointest. Surg.*, 2000, **4**, 50–54.

71. R. Snepar, G. A. Poporad, J. M. Romano, W. D. Kobasa and D. Kaye, *Infect. Immun.*, 1982, **36**, 518–524.

72. W. S. Ruddell, A. T. Axon, J. M. Findlay, B. A. Bartholomew and M. J. Hill, *Lancet*, 1980, **1**, 672–674.

73. B. K. Sharma, I. A. Santana, E. C. Wood, R. P. Walt, M. Pereira, P. Noone, P. L. Smith, C. L. Walters and R. E. Pounder, *Br. Med. J. (Clin. Res. Ed)*, 1984, **289**, 717–719.

74. E. Verdu, F. Viani, D. Armstrong, R. Fraser, H. H. Siegrist, B. Pignatelli, J. P. Idstrom, C. Cederberg, A. L. Blum and M. Fried, *Gut*, 1994, **35**, 455–460.

75. Y. Karmeli, R. Stalnikowitz, R. Eliakim and G. Rahav, *Dig. Dis. Sci.*, 1995, **40**, 2070–2073.

76. J. Thorens, F. Froehlich, W. Schwizer, E. Saraga, J. Bille, K. Gyr, P. Duroux, M. Nicolet, B. Pignatelli, A. L. Blum, J. J. Gonvers and M. Fried, *Gut*, 1996, **39**, 54–59.

77. K. Shindo, R. Yamazaki, K. Koide, M. Fukumura and Y. Hirai, *J. Investig. Med.*, 1996, **44**, 462–469.

78. K. Shindo, M. Machida, M. Fukumura, K. Koide and R. Yamazaki, *Gut*, 1998, **42**, 266–271.

79. A. Roda and A. Fini, *Hepatology*, 1984, **4**, 72S–76S.

80. J. W. Harmon, L. F. Johnson and C. L. Maydonovitch, *Dig. Dis. Sci.*, 1981, **26**, 65–72.

81. P. Hrelia, C. Fimognari, F. Maffei, G. Brandi, G. Biasco and G. Cantelli-Forti, *Mutat. Res.*, 2002, **514**, 125–132.

82. L. A. Feagins, H. Y. Zhang, K. Hormi-Carver, M. H. Quinones, D. Thomas, X. Zhang, L. S. Terada, S. J. Spechler, R. D. Ramirez and R. F. Souza, *Am. J. Gastroenterol.*, 2007, **102**, 10–20.

83. S. R. DeMeester and T. R. DeMeester, *Ann. Surg.*, 2000, **231**, 303–321.

84. E. J. Schweitzer, B. L. Bass, S. Batzri and J. W. Harmon, *Dig. Dis. Sci.*, 1986, **31**, 1105–1113.

85. S. Batzri, J. W. Harmon, E. J. Schweitzer and R. Toles, *Proc. Soc. Exp. Biol. Med.*, 1991, **197**, 393–399.

86. R. E. Marshall, A. Anggiansah, D. K. Manifold, W. A. Owen and W. J. Owen, *Gut*, 1998, **43**, 603–606.

87. A. Wilmer, J. Tack, E. Frans, H. Dits, S. Vanderschueren, A. Gevers and H. Bobbaers, *Gastroenterology*, 1999, **116**, 1293–1299.

88. H. J. Stein, W. K. Kauer, H. Feussner and J. R. Siewert, *J. Gastrointest. Surg.*, 1998, **2**, 333–341.

89. A. I. Sarela, D. G. Hick, C. S. Verbeke, J. F. Casey, P. J. Guillou and G. W. Clark, *Arch. Surg.*, 2004, **139**, 547–551.

90. J. A. Todd, K. K. Basu and J. S. de Caestecker, *Aliment. Pharmacol. Therap.*, 2005, **21**, 969–975.

91. N. J. Shaheen, *Am. J. Gastroenterol.*, 2005, **100**, 1009–1011.

92. W. L. Hofstetter, J. H. Peters, T. R. DeMeester, J. A. Hagen, S. R. DeMeester, P. F. Crookes, P. Tsai, F. Banki and C. G. Bremner, *Ann. Surg.*, 2001, **234**, 532–538; discussion 538–539.
93. J. Freedman, W. Ye, E. Naslund and J. Lagergren, *Gastroenterology*, 2001, **121**, 548–553.
94. C. S. Neumann, T. H. Iqbal and B. T. Cooper, *Aliment. Pharmacol. Ther.*, 1995, **9**, 451–454.
95. H. B. El-Serag, T. V. Aguirre, S. Davis, M. Kuebeler, A. Bhattacharyya and R. E. Sampliner, *Am. J. Gastroenterol.*, 2004, **99**, 1877–1883.
96. S. R. DeMeester, *Semin. Thorac. Cardiovasc. Surg.*, 1997, **9**, 279–284.
97. S. E. Attwood, T. R. DeMeester, C. G. Bremner, A. P. Barlow and R. A. Hinder, *Surgery*, 1989, **106**, 764–770.
98. W. K. Kauer, J. H. Peters, T. R. DeMeester, A. P. Ireland, C. G. Bremner and J. A. Hagen, *Ann. Surg.*, 1995, **222**, 525–531; discussion 531–523.
99. G. Marcus, *N. Engl. J. Med.*, 1999, **341**, 537; author reply, 537–538.
100. J. Lagergren, R. Bergstrom, A. Lindgren and O. Nyren, *N. Engl. J. Med.*, 1999, **340**, 825–831.
101. L. A. Garcia Rodriguez, J. Lagergren and M. Lindblad, *Gut*, 2006, **55**, 1538–1544.
102. F. Fioretti, A. Tavani, C. La Vecchia and S. Franceschi, *Eur. J. Cancer Prevent.*, 1997, **6**, 143–146.
103. D. G. Colin-Jones, M. J. Langman, D. H. Lawson, R. F. Logan, K. R. Paterson and M. P. Vessey, *Q. J. Med.*, 1991, **78**, 13–19.
104. H. Moller, K. Lindvig, R. Klefter, J. Mosbech and O. Moller Jensen, *Gut*, 1989, **30**, 1558–1562.
105. C. R. Haigh, S. E. Attwood, D. G. Thompson, J. A. Jankowski, C. M. Kirton, D. M. Pritchard, A. Varro and R. Dimaline, *Gastroenterology*, 2003, **124**, 615–625.
106. S. I. Abdalla, P. Lao-Sirieix, M. R. Novelli, L. B. Lovat, I. R. Sanderson and R. C. Fitzgerald, *Clin. Cancer Res.*, 2004, **10**, 4784–4792.
107. E. B. Henry, A. Carswell, A. Wirz, V. Fyffe and K. E. McColl, *Aliment. Pharmacol. Ther.*, 2005, **22**, 539–545.
108. V. A. Luketic and A. J. Sanyal, *Gastroenterologist*, 1994, **2**, 74–79.
109. I. Makino and H. Tanaka, *J. Gastroenterol. Hepatol.*, 1998, **13**, 659–664.
110. M. E. Gershwin, A. A. Ansari, I. R. Mackay, Y. Nakanuma, A. Nishio, M. J. Rowley and R. L. Coppel, *Immunol. Rev.*, 2000, **174**, 210–225.
111. E. J. Heathcote, *Hepatology*, 2000, **31**, 1005–1013.
112. B. Y. Tung, M. J. Emond, R. C. Haggitt, M. P. Bronner, M. B. Kimmey, K. V. Kowdley and T. A. Brentnall, *Ann. Intern. Med.*, 2001, **134**, 89–95.
113. S. A. Shah, Y. Volkov, Q. Arfin, M. M. Abdel-Latif and D. Kelleher, *Int. J. Cancer*, 2006, **118**, 532–539.
114. S. S. Joo, T. J. Won and D. I. Lee, *Arch. Pharm. Res.*, 2004, **27**, 954–960.
115. M. Fein, J. Maroske and K. H. Fuchs, *Brit. J. Surg.*, 2006, **93**, 1475–1482.
116. H. J. Stein, H. Feussner, W. Kauer, T. R. DeMeester and J. R. Siewert, *Am. J. Surg.*, 1994, **167**, 163–168.

CHAPTER 7
Bile Acids and Obesity

LUCINDA SUMMERS[1] AND LAURA J. HARDIE[2*]

[1] Department of Cardiovascular Medicine, Academic Unit of Molecular Vascular Medicine; [2] Molecular Epidemiology Unit, Leeds Institute for Genetics, Health and Therapeutics, LIGHT Laboratories, Clarendon Way, University of Leeds, Leeds LS2 9JT, UK

7.1 Obesity

Obesity is a complex chronic condition with genetic, environmental, metabolic and psychological aspects. It has biochemical, biomechanical and sociological effects that lead to increased ill-health and early death and represents one of the biggest healthcare challenges of the 21st century.

The World Health Organisation (WHO) defines obesity as a body mass index (BMI) greater than or equal to $30 \, kg/m^2$, where BMI is calculated as weight in kilogram divided by the square of height in metre.[1] Table 7.1 shows the WHO classification in full. BMI is a simple index of obesity and correlates well with total adiposity,[2] morbidity and mortality from many diseases.[3]

This relationship is age independent and the same for both sexes, but may not correspond to the same degree of fatness across different populations, partly due to differences in body proportions. Many would therefore advocate the use of waist circumference and waist:hip ratio measurements in conjunction with BMI, as these measurements are more closely related to abdominal fat accumulation, which is implicated in the development of type-2 diabetes and heart disease.

---

[*]corresponding author

Issues in Toxicology
Bile Acids: Toxicology and Bioactivity
Edited by Gareth Jenkins and Laura J. Hardie
© Royal Society of Chemistry, 2008

Table 7.1 World Health Organisation classification of obesity.[1]

Classification	Body mass index (kg/m^2)	Risk of co-morbidities
Underweight	< 18.5	Low (but risk of other clinical problems increased)
Normal Range	18.5–24.9	Average
Overweight		
Pre-Obese	25–29.9	Increased
Obese Class I	30.0–34.9	Moderate
Obese Class II	35.0–39.9	Severe
Obese Class III	≥ 40.0	Very severe

7.1.1 Epidemiology

Obesity is a global epidemic,[1] with over 1 billion overweight adults worldwide, at least 300 million of whom are obese. People with a healthy BMI, as defined by WHO, are now in a minority in the USA and Western Europe. In England, 8% of women and 6% of men were classified as obese in 1980.[4] By 1998, the prevalence of obesity had virtually trebled to 21% of women and 17% of men.[4] In 2007, over 50% of women and 66% of men in England are either overweight or obese.[5] Even in developing countries, overweight and obesity are becoming major causes of ill-health, ahead of the traditional problems of malnutrition and infectious diseases. In 2000, it was estimated that over 115 million people in developing countries suffer from obesity-related problems (WHO 2000) and this figure continues to increase.

According to WHO, the prevalence of adult obesity varies from 0.7% in India (data from 1998) to 78.5% in Nauru (data from 1994). The prevalence of adult obesity in India is predicted to have increased dramatically over the 13 years since these data were collected, as it has in other developing countries: in a study of 7 provinces in China, overweight doubled in women and nearly tripled in men between 1989 and 1997.[6] Some populations have a particularly high prevalence of obesity, for example, Pacific Islanders (Nauruans) and North American Indians (Pima Indians) and within a given country different racial groups are more prone to obesity than others: in the USA, the prevalence of overweight and obesity is higher in Hispanic men than in non-Hispanic white or black men and is higher in both black and Hispanic women than in non-Hispanic white women.[7] Whilst some people are genetically pre-disposed to weight gain, the main reason for the rapid increase in obesity is likely to be a combination of altered dietary habits and less active lifestyles.

Obesity increases steadily with age in the under 60s.[8] However, obesity is affecting younger people more than ever before. In the USA, there was an increase in prevalence of overweight (at or above the 95th percentile of sex-specific BMI for age growth charts) in female children and adolescents from 13.8% in 1999–2000 to 16.0% in 2003–2004 and an increase in male children and adolescents from 14.0% to 18.2% in the same time period.[9] Similarly, in the UK, although the figures are lower, the prevalence of obesity in 5–10 year

Table 7.2   Obesity-related cancers.[11]

BMI-related cancers	*Non-BMI related cancers*
Oesophageal and stomach (men only)	Oesophageal and stomach (women only)
Colorectal	Melanoma
Liver	Bladder
Gallbladder	Brain
Pancreatic	Leukaemia (women only)
Lung	
Breast (women only)	
Cervical and uterine	
Ovarian	
Prostate	
Kidney	
Non-Hodgkin's lymphoma	
Multiple myeloma	
Leukaemia (men only)	

olds increased from 1.2% in 1984 to 3.4% in 1996–7 and 6.0% in 2002–3 in boys and from 1.8% in 1984 to 4.5% in 1996–7 and 6.6% in 2002–3 in girls.[10]

7.1.2   Morbidity and Mortality

As the prevalence of obesity increases worldwide, so does the prevalence of associated co-morbidities: type-2 diabetes, chronic obstructive sleep apnoea, cardiovascular disease (hyper-tension, coronary artery disease and congestive heart failure, stroke and peripheral vascular disease), fatty liver disease, various malignancies (Table 7.2), gallstones, subfertility, musculo-skeletal problems and depression.

In a prospective study of over 900,000 adults in the USA, increased BMI was associated with increased death rates from all cancers combined, as well as for cancers at multiple specific sites.[11] The authors of this study estimated that obesity and overweight in the USA could account for 14% of all deaths from cancer in men and 20% in women. As well as the substantial human cost from obesity-associated disease and premature mortality, obesity has serious consequences for the economy. In the UK, it has been estimated that obesity leads to 18 million sick days a year, 40,000 lost years of working life and shortening of life by 9 years on average. Treatment costs to the National Health Service are of the order of £0.5 billion a year and the impact on the wider economy may be as high as £2 billion.[5]

7.2   Adipose Tissue

As people become obese they accumulate more fat, but not all fat cells (adipocytes) are equal: there are different types of adipose tissue and adipocytes behave differently depending on the location of the fat depot.

The two main types of fat are white and brown adipose tissue (WAT and BAT respectively). WAT normally makes up about 10–30% of adult human body weight, although this can increase to 60% in the morbidly obese. Adipocytes from WAT store lipid in a unilocular droplet contrasting with adipocytes from BAT that have multi-locular lipid droplets.[12] BAT forms about 5% of body weight in human infants but this tissue has virtually disappeared by adulthood. BAT is highly vascularised and the adipocytes contain high numbers of mito-chondria, giving the tissue its brown appearance. In addition to lipid storage, BAT plays a pivotal role in nonshivering thermogenesis; expression of mito-chondrial uncoupling protein 1 in these adipocytes permits uncoupling of oxi-dative phosphorylation in the mitochondria and the generation of heat.[13]

For at least 50 years it has been known that being "pear-shaped", *i.e.* accu-mulating fat on the buttocks and thighs, as women have traditionally tended to do, is protective against diabetes and heart disease, whereas being "apple-shaped", *i.e.* storing fat in the abdominal region, as men do, is associated with an increased risk of diabetes and heart disease.[14] Central obesity, the apple-shape, is characterised by increased intra-abdominal visceral fat storage around the ab-dominal organs, whilst lower-body obesity, the pear-shape, is characterised by increased fat storage in the subcutaneous fat depots of the hips, thighs and lower trunk. Visceral fat has been specifically linked with metabolic disturbances and it has been suggested that this is because of the release of fatty acids from this fat depot directly to the liver.[15] However, there are also associations between sub-cutaneous fat on the trunk and an adverse metabolic profile,[16,17] perhaps as the result of the production by adipose tissue of hormones (see below) that have effects on metabolism, since, although subcutaneous fat makes up 80% of total body fat, venous blood from subcutaneous fat depots does not drain directly into the liver. A more recent finding is that excess fat stored in the liver[18] and skeletal muscle[17] is linked with metabolic abnormalities. This fat is referred to as "ec-topic" fat and may pre-dispose to type-2 diabetes, although it is possible that it occurs as a result of metabolic changes rather than causing them.

7.2.1 Regulation of Fatty Acid and Glucose Metabolism in Adipocytes

In the fed state, fatty acids and glucose are actively taken up by adipocytes to produce triglyceride that is stored in lipid droplets (Figure 7.1). In the "starved" state, the post-absorptive period, the triglyceride reserve is broken down to re-lease fatty acids (and glycerol) into the circulation to provide energy for skeletal and cardiac muscle. These events are tightly regulated by insulin (Figure 7.2).

Following feeding, insulin is secreted by β-cells in the pancreatic islets in response to increased concentrations of glucose and amino acids, the release of gut hormones (*e.g.* glucagon-like peptide-1 (GLP-1), gastric inhibitory poly-peptide (GIP) and stimulation by the vagus nerve. Insulin suppresses glucose production by the liver and stimulates glucose uptake by skeletal muscle and fat tissue, thereby decreasing circulating glucose concentrations. Insulin also

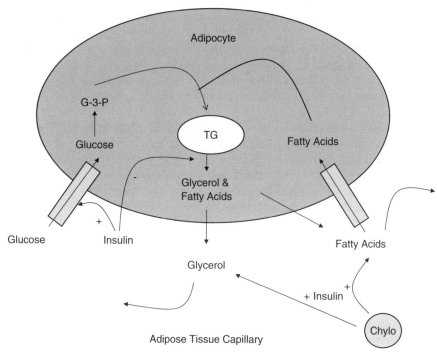

Figure 7.1 Regulation of triglyceride and fatty-acid metabolism in adipose tissue. In the starved state, the adipocyte triglyceride (TG) store is broken down into fatty acids and glycerol that are released by the cell. Fatty acids can be taken up again by the adipocyte, but most will be released into the circulation to serve as an energy supply for other tissues. In the fed state, insulin suppresses this process and stimulates the breakdown of the triglyceride stored in circulating chylomicrons (Chylo). Fatty acids released are, for the most part, taken up by the adipocyte and re-esterified to form triglyceride that is stored. Insulin stimulates the re-esterification of fatty acids and also the active transport of glucose into the cell, which is converted into glycerol-3-phosphate (G-3-P) for re-esterification with the fatty acids.

suppresses the release of fatty acids from adipocytes and stimulates the breakdown of triglyceride from circulating chylomicrons. The resulting fatty acids released from the chylomicrons are taken up into the adipocyte and re-esterified to form triglyceride stores. Insulin also stimulates the active transport of glucose into the cell, which is converted into glycerol-3-phosphate (G-3-P) for re-esterification with the fatty acids.

7.2.2 Adipose Tissue as an Endocrine Organ

In addition to its pivotal role in fat and glucose metabolism, adipose tissue is increasingly recognised as a tissue with significant endocrine and other

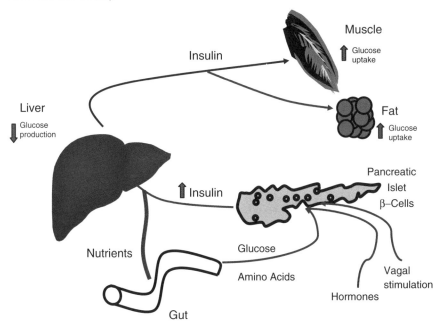

Figure 7.2 Glucose regulation by insulin. Increased concentrations of glucose and amino acids in the circulation, vagus nerve stimulation and circulating hormones (*e.g.* GLP) stimulate insulin secretion by β-cells of the pancreatic islets. Insulin suppresses glucose production by the liver and increases glucose uptake by skeletal muscle and fat, leading to decreased circulating glucose concentrations.

functions. Adipocytes secrete a wide range of factors including hormones, cytokines, complement factors and enzymes (Table 7.3). These factors regulate important and diverse functions such as appetite and energy balance, inflammation and immune responses, and the remodelling of blood vessels (Table 7.3).

7.3 Common Complications of Obesity

As people increase in adiposity, their biochemistry tends to alter in a characteristic way that can lead to clinical complications.

7.3.1 Insulin Resistance

Insulin resistance (IR) was originally defined by Berson and Yalow (quoted in[19]) as a state (of a cell, tissue, system or body) in which greater than normal amounts of insulin are required to elicit a quantitatively normal response. It is said to be present when the ability of insulin to stimulate the uptake and disposal of glucose

Table 7.3 Examples of factors secreted by adipocytes and their modulatory function.

Function	Adipocyte-derived factor
Appetite and energy homeostasis	leptin, neuropeptide Y, adiponectin, resistin, visfatin, interleukin-6, retinol binding protein, cortisol
Acute phase reactants and immune system	tumour necrosis factor-α, interleukin-6 and -8, complement factors C3, B and D, monocyte chemoattractant protein, cortisol, serum amyloid A3, lipocalin 24p3, C reactive protein
Lipid and lipoprotein metabolism	acylation stimulating protein, lipoprotein lipase, cholesteryl ester transfer protein
Haemostasis	plasminogen activator inhibitor-1
Vascular function and angiogenesis	vascular endothelial growth factor, fasting-induced adipose factor, angiopoietin-2, angiotensinogen, angiotensin II, prostaglandins heparin binding epidermal growth factor-like growth factor
Extracellular matrix remodelling	Type-VI collagen, plasminogen activator inhibitor-1, metalloproteases and their tissue inhibitors

Table 7.4 The cluster of abnormalities that make up the insulin-resistant syndrome.

Features of insulin resistance
Glucose intolerance
↑ Plasma TG
↓ HDL-cholesterol
Smaller, denser LDL
Hyper-tension
Hyper-uricaemia
↑ PAI-1
↑ C-reactive protein
Microalbuminuria
↑ Homocysteine

TG, triglyceride; HDL, high-density lipoprotein; LDL, low-density lipoprotein; PAI-1, plasminogen activator inhibitor-1. Note: not all of these features will be present in every insulin-resistant individual.

is impaired.[20] The incidence of insulin resistance increases with increasing BMI, making insulin resistance a common feature of obesity. Insulin resistance is also associated with a cluster of cardiovascular risk factors (Table 7.4) and this has been variously termed the insulin-resistance syndrome, syndrome X, Reaven's syndrome and, more recently, metabolic or cardiometabolic syndrome. The underlying cause of insulin resistance and associated cardiovascular risk factors remains unclear.

7.3.2 Impaired Glucose Tolerance

As mentioned above, insulin secretion by β-cells of the pancreatic islets increases in response to increasing glucose concentrations. In the insulin-resistant state, despite insulin concentrations that are increased two- to three-fold, there is an excessive rate of liver-glucose production. In addition, skeletal muscle glucose disposal in response to insulin is markedly decreased. This results in increased glucose concentrations. This inability to control glucose concentrations, is referred to as "impaired glucose tolerance" and may ultimately lead to type-2 diabetes.

7.3.3 Dyslipidemia

Lipoproteins are an important class of serum proteins in which a spherical hydrophobic core of triglycerides or cholesterol esters is surrounded by an amphipathic monolayer of phospholipids, cholesterol and apolipoproteins (fat-binding proteins). Lipoproteins transport lipid in the circulation and vary in size and density, depending on their protein:lipid ratio (Figure 7.3). Lipoprotein metabolism is adversely affected by obesity; low-density lipoprotein (LDL)-cholesterol and plasma triglyceride are increased, together with decreased high-density lipoprotein (HDL)-cholesterol concentrations.

As discussed above, insulin suppresses the breakdown of triglyceride within fat cells in the post-prandial period, preventing release of fatty acids from adipocytes in healthy individuals. Insulin also stimulates triglyceride clearance from triglyceride-rich lipoprotein particles and the esterification of fatty acids to form the intra-adipocyte triglyceride store.

As a result of insulin resistance, insulin fails to suppress the release of fatty acids from the intra-adipocyte fat store in the post-prandial period, resulting in increased fatty-acid availability. This, in turn, results in increased production of triglyceride-rich, very low density lipoprotein (VLDL) by the liver. There is also decreased clearance of triglyceride-rich lipoproteins, due to the failure of insulin to stimulate triglyceride clearance in the post-prandial period. This results in increased triglyceride-cholesteryl ester exchange between the cholesterol-rich HDL and LDL and the triglyceride-rich VLDL and chylomicron by cholesteryl ester transfer protein. Relatively cholesterol-depleted HDL and LDL are produced as a result of these exchanges, together with a reduction in their size due to the breakdown of their comparatively triglyceride-rich cores by a liver enzyme, hepatic lipase. Insulin resistance is therefore associated with a particularly adverse lipid profile: small, dense LDL particles, cholesterol-depleted HDL particles and increased concentrations of triglyceride-rich lipoprotein remnants. This type of dyslipidemia is strongly causally associated with the development of cardiovascular disease.

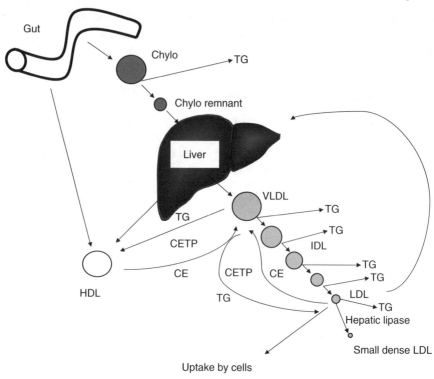

Figure 7.3   The metabolism of lipoprotein particles. After a meal, digested fat is
packaged by the gut as triglyceride (TG) within chylomicrons (Chylo). As
these particles circulate, the triglyceride is gradually removed, leaving
chylomicron remnants that are taken up by the liver. The liver releases
triglyceride-rich, very low-density lipoproteins (VLDL), from which triglyceride is removed in a similar fashion, producing intermediate-density
lipoproteins (IDL) and, finally, low-density lipoproteins (LDL) that are
taken up by the liver or peripheral cells. High-density lipoprotein (HDL)
is secreted both by the gut and the liver. Cholesteryl ester transfer protein
(CETP) transfers cholesteryl ester (CE) from HDL and LDL to the triglyceride-rich particles and triglyceride in the other direction. This process, and the further breakdown of HDL and LDL by hepatic lipase,
results in the formation of cholesterol-depleted HDL, small dense LDL
and triglyceride-rich remnant particles, which are all implicated in the
development of cardiovascular disease.

7.4   Effect of Bile Acids on Obesity

In 1997, a study by Ikemoto and colleagues[21] was the first to suggest a link for
bile acids in the regulation of obesity. They demonstrated that addition of the
common bile salt, sodium cholate (0.5%) to a high-fat diet prevented increases
in total body weight and WAT in C57BL/6 mice compared with littermates fed
an unsupplemented high-fat diet. These effects occurred despite similar energy
intakes across control (high carbohydrate), high-fat and cholate-supplemented

high-fat diet groups, indicating that the cholate-associated effect was not simply a reflection of reduced calorie intake in the supplemented animals. Cholate supplementation also prevented the development of hyper-glycemia and reduced blood insulin levels compared with high-fat-fed animals. Similar effects were reported with supplementation of the bile acid, chenodeoxycholate acid.

Feeding high-fat diet was also associated with increased triglyceride levels in liver and gastrocnemius; effects that were completely prevented by the addition of cholate to the diet. In order to establish the possible mechanism underlying the reduction in triglyceride levels, the authors measured acyl-CoA synthetase in the liver of these animals. This enzyme catalyses the formation of acyl-CoA, which is utilised in triglyceride generation in the liver. Levels of this enzyme were reduced by 68% in the liver of cholate-supplemented high-fat-fed mice compared with high-fat-fed controls.

In a landmark study, Watanabe and colleagues[22] have recently confirmed and extended these findings, linking bile acids directly to the regulation of thermogenesis.

The authors demonstrated that not only does cholic-acid-supplementation reduce high-fat-diet-induced weight gain but it also reverses weight gain once established in obese mice, resulting in reduced WAT mass and improved glucose tolerance. Once more these effects were not a result of reduced feeding and lower calorific intake in these cholic-acid-supplemented animals. Rather, indirect calorimetry demonstrated that animals fed a high-fat diet containing cholic acid had higher CO_2 production and O_2 consumption compared with control animals fed a high-fat diet or regular diet, indicating increased energy expenditure in these animals. Electron microscopy revealed that cholic-acid supplementation did not cause the adipocyte hyper-trophy observed in animals on high-fat diet alone, and that in BAT, mitochondria of cholic-acid-fed animals had increased numbers of lamellar cristae, implicating BAT in the cholic-acid-induced increased energy expenditure. Microarray analysis of BAT revealed that mRNA for the enzyme, type-2 iodothyronine deiodinase (D2) was strongly up-regulated in cholic-acid-fed mice. Up-regulation of this gene also occurred in BAT cells treated with bile acids *in vitro*. D2 converts thyroxine (T4) to active 3,5,3'-tri-iodothyronine (T3) and is required in BAT for adaptive thermogenesis.[23] The authors demonstrated that bile acids caused cAMP-PKA activation in BAT cells and this was mediated via the G-protein-coupled receptor, TGR5 (Figure 7.4). This receptor is ubiquitously expressed in human tissues and the following bile acids stimulate signal transduction; lithocholic acid > deoxycholic acid > chenodeoxycholic acid > cholic acid.[24,25]

Contrasting with rodents, BAT is found in small amounts in adult humans. It has been proposed that skeletal muscle rather than BAT may play a pivotal role in energy homeostasis in adult humans. The authors also demonstrated that cultured human skeletal muscle myoblasts express D2 and high levels of TGR5, and a number of common bile acids (cholic acid, taurocholic acid, deoxycholic acid, chenodeoxycholic acid) were able to increase cAMP levels concomitant with increased D2 activity[22] (Figure 7.4). Taurocholic acid was also able to

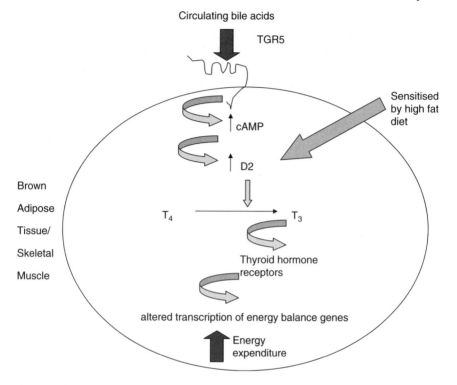

Figure 7.4   The effect of bile acids on energy expenditure. Circulating bile acids bind
to the G-protein-coupled receptor, TGR5 that stimulates increased
cAMP-PKA activation and increased expression of type-2 iodothyronine
deiodinase (D2). This response is sensitised by a high-fat diet. D2 converts
thyroxine (T4) to active 3,5,3′-tri-iodothyronine (T3). T3 stimulates
thyroid hormone receptor binding to target genes. This leads to altered
expression of genes associated with energy balance, and increased energy
expenditure.

directly increase aerobic mitochondrial oxidation in these cells as detected
through increased oxygen consumption.

These results are exciting and together suggest that direct targeting of the
TGR5 receptor could provide a novel route to modulating metabolic rate and
obesity.

7.4.1   Bile Acids, the Farnesoid X Receptor (FXR) and Fat Metabolism

Bile acids are also natural ligands for the farnesoid X receptor (FXR), a
receptor that belongs to the nuclear hormone receptor superfamily.[26–28]
The hydrophobic bile acid chenodeoxycholic acid (CDCA) is the most potent

bile-acid activator of FXR, whereas hydrophilic ursodeoxycholic acid (UDCA) demonstrates little activity.[26–28] Until recently the principal role of FXR was considered to protect liver cells from high levels of bile acids by decreasing endogenous production of bile acids and accelerating their transformation and excretion (for a full review see ref. 29). However, in addition to this function, there is accumulating evidence that FXR has direct effects on adipocyte differentiation and function and may play an important role in the regulation of systemic lipid metabolism and peripheral glucose homeostasis.[29]

In vivo, FXR is expressed in white adipose tissue (WAT) with levels decreased in the WAT of dietary (high fat) and genetic (*ob*/*ob* mice) models of obesity.[30]

Transactivation of FXR by chenodeoxycholic acid has been shown to enhance adipogenesis in 3T3-L1 cells as measured by increased C\EBP, PPAR-γ2, FABP, SREBP-1c and AdipoQ expression.[31] The functional significance of FXR expression in adipocytes was elegantly demonstrated by Cariou *et al.*[30] who showed that direct agonism of FXR with the compound GW4064 increases insulin signalling and insulin-stimulated glucose uptake in adipocytes. Loss of FXR using a gene knockout model resulted in impaired adipocyte differentiation, characterised by reduced aP2, PPPARγ, C\EBPα, leptin and GLUT4 expression. Furthermore lipid storage was compromised; murine embryonic fibroblasts (MEFs) from FXR knockout mice showed poor accumulation of triglyceride. These *in-vitro* results were consistent with the *in-vivo* phenotype; FXR knockout mice have reduced adipocyte size, adipose tissue mass, and lower circulating leptin concentrations.

With respect to glycaemic parameters FXR-deficient mice ($FXR^{-/-}$) show impaired glucose tolerance, whole-body glucose disposal and insulin resistance. FXR deficiency is also associated with impaired distal insulin signalling in muscle and WAT through reduced insulin-dependent AKT phosphorylation.[30]

Finally, dyslipidemia is a feature of FXR knockout mice; levels of VLDL, LDL and HDL cholesterol and ApoB lipoproteins are increased concomitant with reduced clearance of HDL cholesteryl ester and elevated triglyceride and free fatty-acid levels.[30–32]

These data raise the possibility that targetted activation of FXR may also be a useful route to manage various aspects of the metabolic phenotype including type-2 diabetes, dyslipidemia and adipocyte function.

7.5 Bile-Acid Sequestrants

As discussed above, obesity is associated with dyslipidemia, a condition where high levels of low-density lipoprotein cholesterol (LDL-C) is common. Elevated LDL-C is strongly associated with an elevated risk of coronary artery disease and for this reason a number of lipid-lowering therapies that target LDL-C have been developed. These include bile-acid sequestrants (BAS), statins (HMG-CoA reductase inhibitors), cholesterol absorption inhibitors, and fibrates.[33]

Bile-acid sequestrants are indigestible, positively charged resins that bind negatively charged bile acids in the lumen of the intestine (reviewed in ref. 34).

Bile acids bound to the resin are no longer available for re-uptake from the large intestine as part of the normal enterohepatic circulation, and are subsequently lost in the faeces. In the region of 30–40% of the endogenous bile-acid pool can be depleted through this route.[35] Reducing bile-acid levels stimulates up-regulation of hepatic CYP7A1 and LDL receptors and increases the catabolism of cholesterol to bile acids. This results in a reduction of cholesterol and associated LDL cholesterol concentrations in the circulation (Table 7.5).[35] In addition to monotherapy, BAS are often used in combination with statins, or nonstatin therapies (fibrates, niacin or cholesterol-absorption inhibitors) to improve lipid- and lipoprotein-lowering effects.

7.5.1   Chemical Composition of BAS

Currently available BAS include cholestyramine, colestipol and colesevelam hydrochloride (colestimide).[34,36,37] Cholestyramine comprises a long-chain polymer of styrene with divinylbenzene trimethylbenzylammonium groups, whereas colestipol is a long-chain polymer of 1-chloro-2,3-epoxypropane with diethylenetriamine. Colesevelam HCl is poly(allylamine hydrochloride) cross-linked with epichlorohydrin and alkylated with 1-bromodecane and 6-bromo-hexyl-trimethylammonium bromide. Bile-acid binding is enhanced and stabilised in the latter compound by long hydrophobic sidechains, increased density of primary amines, and quaternary amine sidechains. For this reason, colesevelam HCl exhibits increased affinity, specificity and capacity to bind bile acids compared with the other BAS.[38] Colesevelam HCl also binds dihydroxy and trihydroxy bile acids with equal affinity, contrasting with cholestyramine and colestipol that preferentially bind dihydroxy bile acids (CDCA and deoxycholic acid).[37–39] The latter BAS can lead to an imbalance towards trihydroxy bile acids and a more hydrophilic bile-acid pool.

7.5.2   Effect of Bile-Acid Sequestrants on Type-2 Diabetes and Obesity

In addition to the protective effects of BAS on atherosclerotic plaques and coronary artery disease,[44–46] a recent study has suggested that BAS could have utility in the management of type-2 diabetes and obesity.

Kobayashi *et al.*[47] supplemented the feed of NSY mice (a model of type-II diabetes) with colestimide. Dietary colestimide supplementation reduced bodyweight gain and BMI when compared with littermates fed high-fat diet alone. This effect was not attributable to a reduction in food consumption, as mice fed colestimide-supplemented high-fat diet actually consumed more food than unsupplemented controls. Non-fasting and fasting blood glucose levels were also significantly reduced in colestimide fed mice compared with high-fat diet controls, indicating better glycaemic control (Table 7.6). Regarding lipid and lipoprotein profiles; triglyceride, cholesterol, total LDL, VLDL and LDL

Table 7.5 The effect of bile-acid sequestrant monotherapy on circulating lipid and lipoprotein profiles.

Trial	Patient group treatment change from baseline (%)	Total Cholesterol	LDL Cholesterol	HDL Cholesterol	Triglyceride
The Type II Coronary Intervention Study Levy et al. (1984)[40]	Cholestyramine (24 g/day) for 5 years	−17	−26	8	28
	placebo	−1	−5	2	26
Superko et al. (1992)[41]	Colestipol (15 g/day) for 12 weeks	−17	−27	1	22
	placebo	1	−2	1.5	13
Insull et al. (2001)[42]	Colesevelam (4.5 g/day) for 24 weeks	−10	−18	3	9
	placebo	1	0	−1	5
Knapp et al. (2001)[43]	Colesevelam (3.8 g/day) for 6 weeks	−9	−16	2	11
	placebo	−2	−4	−3	6

Table 7.6 Effect of dietary colestimide on biochemical measures in diabetic NSY mice both during and following high-fat diet induced obesity.[47]

	Effects of colestimide during obesity development	Effects of colestimide on established obesity
Body weight	↓	no effect
BMI	↓	no effect
Non-fasting blood glucose	↓	↓
Fasting blood glucose	↓	N/D
Non-fasting plasma insulin	↓	↓
Fasting plasma insulin	↓	↓
Liver weight	↓	↓
Subcutaneous fat	↓	no effect
Visceral fat	no effect	no effect
Brown adipose tissue	no effect	no effect
Pancreatic islet area	↓	no effect
Fatty liver (Sudan black)	↓	↓
Total cholesterol	↓	no effect
Chylomicron-cholesterol	↓	↓
VLDL-cholesterol	↓	↓
LDL-cholesterol	↓	no effect
HDL-cholesterol	↓	↓
Total triglyceride	↓	no effect
Chylomicron-triglyceride	no effect	↓
VLDL-triglyceride	↓	no effect
LDL-triglyceride	↓	↓
HDL-triglyceride	↓	↓
Glucose tolerance	↓	↓
Insulin sensitivity	↑	↑
Insulin secretion	↓	↓
Faecal lipid excretion	↑	N/D
Faecal bile-acid content	↑	N/D

N/D – not determined.

cholesterol were all significantly reduced in colestimide-supplemented mice when compared with control animals.

Importantly, the authors also examined whether colestimide ameliorated obesity and its associated effects once established. Body weight and BMI were not significantly reduced in mice fed colestimide, but insulin and glucose profiles, serum cholesterol and triglyceride concentrations were all improved compared with high-fat-diet-fed controls.

Faecal bile-acid content was increased in colestimide-supplemented mice consistent with the mode of action of the BAS. In addition, faecal lipid content was raised in treated animals, suggesting reduced lipid absorption, presumably *via* the lowering of duodenal bile-acid concentration.

Changes in expression of key genes involved in lipid and glucose metabolism were associated with the effects of colestimide. mRNA for the cholesterol-catabolising enzyme, CYP7A1, was increased 4-fold in colestimide-fed animals and was coincident with a decrease in the CYP7A1 transcriptional repressor, SHP. The lipogenic gene regulator SREBP-1c, was also reduced in animals fed colestimide, although levels of FXR, the nuclear receptor for bile acids (see above) and a regulator of SHP were unaffected. Genes associated with fatty-acid biosynthesis and β-oxidation were modulated: mRNA levels of both fatty-acid synthase, a gene regulated by SREBP-1c, and the gluceoneogenesis-associated enzyme, PEPCK, were significantly reduced in animals fed colestimide in the prevention study. Similar trends for CYP7A1 and SHP gene expression were observed in the intervention study.

Together these data suggest that colestimide and possibly other BAS may have utility not only in the management of dyslipidemias but also in the control of obesity and type-II diabetes.

If the findings relating to obesity and improved glycaemic control can be confirmed in human studies such drugs would be highly attractive. As discussed above, bile-acid sequestrants have been used for many years to treat dyslipidemia in relation to reducing cardiovascular disease risk and the safety profile of these compounds is well established. However, due to the large doses of compound that require to be consumed, compliance is an issue for BAS therapies. In the future, this may be resolved with the development of more specific and efficient resins that require lower doses.

7.6 Summary

Obesity is a global epidemic that causes significant morbidity and mortality. Traditionally, bile acids have been linked to the development of obesity through their role in the intestinal absorption of fatty acids. However, recent studies indicate that bile acids have a much wider role in the regulation of energy balance in the body. Bile acids are natural ligands for TGR5 and FXR receptors and modulate adipocyte differentiation and function, thermogenesis, glucose, lipid and insulin homeostasis. These findings have recently raised the possibility that dietary bile-acid sequestration and/or targeting of bile-acid signalling pathways may ameliorate obesity and associated pathologies such as insulin resistance, impaired glucose tolerance and dyslipidemia and promise to be highly active areas of research in the future.

References

1. World Health Organisation Obesity: Preventing and managing the global epidemic. 2000. Technical Report Series. Number 894, 265pp epub: http://www.who.int/bookorders/index.htm
2. J. D. Webster, R. Hesp and J. S. Garrow, *Hum. Nutr. Clin. Nutr.*, 1984, **38**, 299.

3. W. C. Willett, W. H. Dietz and G. A. Colditz, *N. Engl. J. Med.*, 1999, **341**, 427.
4. Department of Health. Health Survey for England: Cardiovascular Disease '98. ed. B. Erens and P. Primatesta, 1999. The Stationery Office, Norwich, UK epub:http://www.archive.official_documents.co.uk/document/doh/survey98/hse-00.htm
5. Report by the comptroller auditor general. HC 220 session 2000-1. 2001. The Stationary Office ordered by the House of Commons, London, UK 6pp epub:http://www.nao.org.uk/publications/nao_reports/00-01/0001220es.pdf
6. A. C. Bell, K. Ge and B. M. Popkin, *Int. J. Obes. Relat. Metab. Disord.*, 2001, **25**, 1079.
7. K. M. Flegal, M. D. Carroll, R. J. Kuczmarski and C. L. Johnson, *Int. J. Obes. Relat. Metab. Disord.*, 1998, **22**, 39.
8. Centres for Disease Control and Prevention. National Health and Nutrition Examination Survey. 2007. epub:http://ww.cdc.gov/nchs/data/databriefs/db01.pdf
9. C. L. Ogden, M. D. Carroll, L. R. Curtin, M. A. McDowell, C. J. Tabak and K. M. Flegal, *Jama*, 2006, **295**, 1549.
10. E. Stamatakis, P. Primatesta, S. Chinn, R. Rona and E. Falascheti, *Arch. Dis. Child.*, 2005, **90**, 999.
11. E. E. Calle, C. Rodriguez, K. Walker-Thurmond and M. J. Thun, *N. Engl. J. Med.*, 2003, **348**, 1625.
12. J. K. Sethi and A. J. Vidal-Puig, *J. Lipid. Res.*, 2007, **48**, 1253.
13. K. S. Echtay, *Free Radical Bio. Med.*, 2007, **43**, 1351.
14. J. Vague, *Am. J. Clin. Nutr.*, 1956, **4**, 20.
15. P. Bjorntorp, *Arteriosclerosis*, 1990, **10**, 493.
16. N. Abate, A. Garg, R. M. Peshock, J. Stray-Gundersen and S. M. Grundy, *J. Clin. Invest.*, 1995, **96**, 88.
17. B. H. Goodpaster, F. L. Thaete, J. A. Simmoneau and D. E. Kelley, *Diabetes*, 1997, **46**, 1579.
18. L. Ryysy, A. M. Hakkinen, T. Goto, S. Vehkavaara, J. Westerbacka, J. Halavaara and H. Yki-Jarvinen, *Diabetes*, 2000, **49**, 749.
19. An overview of insulin resistance. Flier, J. S. 1-8. 1993. In Insulin Resistance. Ed D. E. Moller, Wiley, Chichester, UK, p. 440.
20. G. M. Reaven, *Physiol. Rev.*, 1995, **75**, 473.
21. S. Ikemoto, M. Takahashi, N. Tsunoda, K. Maruyama, H. Itakura, K. Kawanaka, I. Tabata, M. Higuchi, T. Tange, T. T. Yamamo and O. Ezaki, *Am. J. Physiol-Endoc. M*, 1997, **36**, E37–E45.
22. M. Watanabe, S. M. Houten, C. Mataki, M. A. Christoffolete, B. W. Kim, H. Sato, N. Messaddeq, J. W. Harney, O. Ezaki, T. Kodama, K. Schoonjans, A. C. Bianco and J. Auwerx, *Nature*, 2006, **439**, 484.
23. A. C. Bianco, D. Salvatore, B. Gereben, M. J. Berry and P. R. Larsen, *Endocr. Rev.*, 2002, **23**, 38.
24. T. Maruyama, Y. Miyamoto, T. Nakamura, Y. Tamai, H. Okada, E. Sugiyama, T. Nakamura, H. Itadani and K. Tanaka, *Biochem. Biophys. Res. Comm.*, 2002, **298**, 714.

25. Y. Kawamata, R. Fujii, M. Hosoya, M. Harada, H. Yoshida, M. Miwa, S. Fukusumi, Y. Habata, T. Itoh, Y. Shintani, S. Hinuma, Y. Fujisawa and M. Fujino, *J. Biol. Chem.*, 2003, **278**, 9435.
26. H. B. Wang, J. Chen, K. Hollister, L. C. Sowers and B. M. Forman, *Mol. Cell.*, 1999, **3**, 543.
27. D. J. Parks, S. G. Blanchard, R. K. Bledsoe, G. Chandra, T. G. Consler, S. A. Kliewer, J. B. Stimmel, T. M. Willson, A. M. Zavacki, D. D. Moore and J. M. Lehmann, *Science*, 1999, **284**, 1365.
28. M. Makishima, A. Y. Okamoto, J. J. Repa, H. Tu, R. M. Learned, A. Luk, M. V. Hull, K. D. Lustig, D. J. Mangelsdorf and B. Shan, *Science*, 1999, **284**, 1362.
29. N. Y. Kalaany and D. J. Mangelsdorf, *Annu. Rev. Physiol.*, 2006, **68**, 159.
30. B. Cariou, K. van Harmelen, D. Duran-Sandoval, T. H. van Dijk, A. Grefhorst, M. Abdelkarim, S. Caron, G. Torpier, J. C. Fruchart, F. J. Gonzalez, F. Kuipers and B. Staels, *J. Biol. Chem.*, 2006, **281**, 11039.
31. G. Rizzo, M. Disante, A. Mencarelli, B. Renga, A. Gioiello, R. Pellicciari and S. Fiorucci, *Mol. Pharmacol.*, 2006, **70**, 1164.
32. C. J. Sinal, M. Tohkin, M. Miyata, J. M. Ward, G. Lambert and F. J. Gonzalez, *Cell*, 2000, **102**, 731.
33. Z. Chilmonczyk, D. Siluk, R. Kaliszan, B. Lozowicka, J. Poplawski and S. Filipek, *Pure Appl. Chem.*, 2001, **73**, 1445.
34. W. Insull, *Southern Med. J.*, 2006, **99**, 257.
35. K. Einarsson, S. Ericsson, S. Ewerth, E. Reihner, M. Rudling, D. Stahlberg and B. Angelin, *Eur. J. Clin. Pharmacol.*, 1991, **40**, S53–S58.
36. Y. Matsuzaki, *J. Gastroen Hepatol.*, 2002, **17**, 1133.
37. K. L. Steinmetz and K. S. Schonder, *Cardiovasc. Drug Rev.*, 2005, **23**, 15.
38. Y. Honda and M. Nakano, *Chem. Pharm. Bull.*, 2000, **48**, 978.
39. G. M. Benson, C. Haynes, S. Blanchard and D. Ellis, *J. Pharm. Sci.*, 1993, **82**, 80.
40. R. I. Levy, J. F. Brensike, S. E. Epstein, S. F. Kelsey, E. R. Passamani, J. M. Richardson, I. K. Loh, N. J. Stone, R. F. Aldrich, J. W. Battaglini, D. J. Moriarty, M. L. Fisher, L. Friedman, W. Friedewald and K. M. Detre, *Circulation*, 1984, **69**, 325.
41. H. R. Superko, P. Greenland, R. A. Manchester, N. A. Andreadis, G. Schectman, N. H. West, D. Hunninghake, W. L. Haskell and J. L. Probstfield, *Am. J. Cardiol.*, 1992, **70**, 135.
42. W. Insull, P. Toth, W. Mullican, D. Hunninghake, S. Burke, J. M. Donovan and M. H. Davidson, *Mayo Clin. Proc.*, 2001, **76**, 971.
43. H. H. Knapp, H. Schrott, P. Ma, R. Knopp, B. Chin, J. M. Gaziano, J. M. Donovan, S. K. Burke and M. H. Davidson, *Am. J. Med.*, 2001, **110**, 352.
44. J. F. Brensike, R. I. Levy, S. F. Kelsey, E. R. Passamani, J. M. Richardson, I. K. Loh, N. J. Stone, R. F. Aldrich, J. W. Battaglini, D. J. Moriarty, M. R. Fisher, L. Friedman, W. Friedewald, K. M. Detre and S. E. Epstein, *Circulation*, 1984, **69**, 313.

45. Anonymous *Arch. Intern. Med.*, 1992, **152**, 1399.
46. M. Studer, M. Briel, B. Leimenstoll, T. R. Glass and H. C. Bucher, *Arch. Intern. Med.*, 2005, **165**, 725.
47. M. Kobayashi, H. Ikegami, T. Fujisawa, K. Nojima, Y. Kawabata, S. Noso, N. Babaya, M. Itoi-Babaya, K. Yamaji, Y. Hiromine, M. Shibata and T. Ogihara, *Diabetes*, 2007, **56**, 239.

CHAPTER 8
The Role of Bile Acids in Cholesterol-Rich Gallstone Formation

LINZI A. THOMAS

Department of Gastroenterology, Singleton Hospital, Swansea NHS Trust, Sketty Lane, Swansea SA2 8QA, UK

8.1 Introduction

Gallstones are a common clinical finding in Western populations, the prevalence rate in Europe being 10–15%.[1] Whilst 80% of cases are asymptomatic from their gallstones, 2% develop complications such as cholecystitis and require surgery.[2]

In industrialised societies, most gallbladder stones are mixed in composition. They contain not only cholesterol but also bile pigments, calcium salts, mucus glycoprotein and often amorphous material. Nonetheless, they are still described as "cholesterol-rich" although the arbitrary definition for cholesterol-rich stones varies from author to author ranging from >70% to >90% by weight.

The majority (approximately 70%) of all cholesterol-rich gallbladder stones (GBS) are radiolucent by plain X-ray and oral cholecystography. However, roughly 50% of stones that are radiolucent by traditional X-ray are visibly dense on CT scanning and have high attenuation scores. The maximum gallstone attenuation score, measured in Hounsfield Units (HU) at the time of CT scanning, provides a noninvasive way of predicting stone composition and dissolvability.[3] Following the advent of laparoscopic cholecystectomy, very few gallstone patients are treated with oral bile acids alone, or in combination with lithotripsy.

Issues in Toxicology
Bile Acids: Toxicology and Bioactivity
Edited by Gareth Jenkins and Laura J. Hardie

8.2 The Triple Defect of Cholesterol GBS Formation

Original research into cholesterol GBS formation was dominated by studies of biliary cholesterol saturation. Bile was obtained either by fine needle puncture of the gallbladder at the time of open cholecystectomy, or by intubating the duodenum of fasting patients, stimulating gallbladder contraction (usually with intravenous cholecystokinin) and aspirating bile-rich fluid. The total bile acid, phospholipids and cholesterol concentrations were then measured, the relative molar percentages of these three lipids derived,[4] and the saturation (or lithogenic) indices calculated.[5]

Results obtained in this way were helpful, but of limited value. The analyses told us whether or not the bile was supersaturated with cholesterol, but did not tell us whether the abnormality was due to too much cholesterol, too few bile acids, too few phospholipids or to some combined defect. The next step, therefore, was to measure the hour-by-hour bile lipid-secretion rates using marker-corrected perfusion techniques. These assume that, in response to the perfusion stimulus (such as an intra-duodenal amino acid mixture), the gallbladder remains tonically contracted throughout and steady-state conditions ensue.

When Reuben *et al.*,[6] measured the hourly secretion rates of phospholipids, bile acids and cholesterol in obese and nonobese individuals with and without cholesterol gallstone disease, they found that the pattern of results was quite different in the obese and the nonobese gallstone carriers. The obese had hyper-secretion of cholesterol but normal bile-acid output, while the nonobese had normal cholesterol secretion but a reduced bile-acid output. The authors speculated that the most likely explanation for the high biliary cholesterol secretion rates in the obese was their increased total body cholesterol synthesis.[7] Conversely, nonobese gallstone carriers often have a reduced total bile-acid pool size,[8] and if the enterohepatic cycling frequency of this small bile-acid pool remains unchanged (controversial), it could explain the reduced bile-acid secretion rate seen in the normal weight (nonobese) individuals.

These results illustrate the concept that there are multiple ways by which bile may become supersaturated with cholesterol. For example, there may be (i) hyper-secretion of cholesterol, (ii) hypo-secretion of bile acids, (iii) hypo-secretion of phospholipids or (iv) some combined secretory defect. Of these, high biliary cholesterol secretion seems to be the most common disorder.

Groups have studied the influence of obesity on gallstone formation.[6,9] Their increased total body cholesterol synthesis is often associated with an increase in the activity of the rate-limiting enzyme in cholesterogenesis – HMG CoA reductase, in the liver. Even more important is the influence of marked weight reduction in the obese as a result of "crash" dieting (very low calorie diets of 500–1000 kcal/day), or of bariatric surgery (usually gastric stapling). During acute weight loss, cholesterol is thought to be mobilised from adipose tissue and secreted, in excess, into bile.

Cholesterol mobilised in this way, as well as that absorbed from the intestine, is taken up by the liver from LDL and chylomicron remnants, *via* receptors on the sinusoidal membranes. Indeed, receptor-mediated cholesterol uptake may be an important primary mechanism for hyper-secretion of biliary cholesterol.

Dietary factors may also be important in gallstone pathogenesis – such as intake of total calories, cholesterol, refined carbohydrates and dietary fibre. As discussed below, the amount of bran, and of other fibre components, in the diet may influence the formation of deoxycholic acid (DCA) in the intestine, its subsequent absorption and its enrichment of bile with DCA conjugates.

Drugs that lower serum lipids, such as clofibrate, may pre-dispose to biliary cholesterol hyper-secretion and to cholesterol gallstones – as may oestrogen-rich oral contraceptives, hormone replacement therapy (HRT) and pregnancy (although the situation in pregnancy is complex since this also affects gall-bladder motility). Diabetes mellitus, particularly the insulin-dependent variety, pre-disposes to cholesterol gallstones as does hyper-triglyceridaemia (but not, surprisingly, hyper-cholesterolaemia).

8.3 Vesicles versus Micelles

A major problem with measurement of saturation indices is that it assumed (incorrectly) that the hydrophobic, water-insoluble cholesterol (CH) was "carried" or transported in mixed micelles with bile acids and phospholipids. However, as a result of the classical studies by Somjen and Gilat,[10] we now know in dilute hepatic bile at least, cholesterol is transported with phospholipids as small, unilamellar (one layered) or multi-lamellar (many layered) vesicles. These consist of bilayers of phospholipids (PLs) that can "carry" cholesterol molecules inter-digitated between the phospholipid molecules. Provided that the molar ratio of CH:PL remains low, the vesicles remain stable with no risk of crystal formation. However, when the CH:PL molar ratio becomes high (around 1:1), the vesicles tend to aggregate and fuse forming large unstable multi-lamellar vesicles. Video-enhanced time-lapse microscopy by Holzbach and others,[11] has shown that it is from these large multi-lamellar vesicles that cholesterol crystals precipitate. However, there is rapid and dynamic exchange of lipid molecules between vesicles, micelles and perhaps other lipid "carriers". It seems likely, therefore, that various crystalline forms of cholesterol (threads, spirals and helices, as well as the classical rhomboid-shaped, notched, triclinic crystals), may precipitate from all the different cholesterol carriers in bile.

8.4 Nucleation of Cholesterol Crystals

When supersaturated fresh gallbladder bile (or model bile) is centrifuged to remove solid crystalline and amorphous precipitates, and supernatant vesicles, and the resultant isotropic (one phase) solution maintained in a dust-free environment at 37 °C and examined daily by light microscopy, cholesterol crystals can be observed to precipitate. The time taken for these solid cholesterol crystals to appear is called the nucleation time (NT).[12] The NT discriminates between stone carriers and stone-free individuals much better than does measurement of saturation indices. Nonetheless, measurement of nucleation time in this way is relatively crude. It represents the cholesterol microcrystal appearance/detection time rather than the true nucleation time. For this reason, a number of crystal-growth assays have been developed that, in essence, are refinements of the original Holan and Holzbach method.

The NT in normal (gallstone free) bile is long ($>$10–14 days) while that in cholesterol gallstone carriers is pathologically rapid (often 1–2 days but always $<$5 days). However, in response to food, the gallbladder normally fills and empties several times per day. Therefore the NT measured *in vitro* does not relate to the residence time of bile within the gallbladder *in vivo*. For this reason, measurements of NT are of relative or comparative value – rather than a physiological measurement.

8.5   Promoters and Inhibitors of Nucleation

In gallstone patients, the nucleation of supersaturated bile requires either an excess of promoters, or a deficiency of inhibitors, of crystallisation (or both). For the past 20 years, attempts have been made to identify these pro- and anti-nucleating agents – but so far without consensus. Suffice to say that the promoters and inhibitors are mainly proteins and that mucous glycoprotein is particularly important – not only as a promoter of nucleation but also because it forms a gel on the surface of the gallbladder mucosa, which is believed to trap cholesterol crystals and contribute to the stasis within the gallbladder. Table 8.1 gives a list of potential promoters and inhibitors.

8.6   Stasis

The third component in the triple defect is stasis of bile within the gallbladder that allows time for the cholesterol crystals to precipitate, agglomerate and

Table 8.1   Proven and putative promoters and inhibitors of nucleation of cholesterol microcrystals in gallbladder bile.

Promoters	*Inhibitors*
Mucin (mucous glycoprotein)	Apolipoprotein A1
Phospholipase C	Apolipoprotein A2
Immunoglobulins (especially lgM/lgA)	Helix pomatia-bound protein
Ionised calcium	
Deoxycholic acid	
Arachadonic acid rich phospholipids	
Low density lipoproteins	
Transferring	

grow to form macroscopic gallstones. Supersaturation of the bile with cholesterol is insufficient to explain crystal precipitation. There must, in addition, be an imbalance between pro- and anti-nucleating factors before precipitation can occur. None of this would be important if the gallbladder contracted vigorously, expelled the bile and flushed the crystals out of the gallbladder, through the cystic bile duct and down the common bile ducts, into the intestine.

There is a large number of studies that suggest that defects in gallbladder contractility contribute, importantly, to the pathogenesis of cholesterol GBS. In brief, many studies have shown that in GBS patients, GB motor function (that is, GB emptying in response to a meal or to an exogenous stimulus such as cholecystokinin – CCK) is impaired when compared with controls. However, there is debate about whether this reduced GB emptying occurs in all gallstone patients or only in a sub-set (strong versus weak "contractors").

8.7   Indirect Evidence that Gallbladder Motor Dysfunction/Stasis is Important in the Pathogenesis of GBS

Table 8.2 summarises some of the conditions in which there is impaired gallbladder motor function and an associated increase in the prevalence of either sludge (which may pre-dispose to stone formation), frank gallstones or both. The role of Octreotide (OT) will now be focused upon since this illustrates the whole sequence of pathogenic changes found in conventional GBS disease – albeit eclipsed into weeks or months.

8.8   The Octreotide Story

Octreotide (OT) is a long-acting somatostatin analogue. When used chronically in the treatment of acromegaly, it suppresses the secretion of growth hormone and insulin-like growth factor -1 (IGF-1). However, its inhibition of peptide hormone release is nonspecific; it also inhibits meal-stimulated CCK release from the intestine. This is the main mechanism for the impaired GB emptying seen during OT treatment that has been implicated in the pathogenesis of Octreotide-induced stones.[13] After 1–2 years treatment, approximately 50% of

Table 8.2   Clinical conditions in which gallbladder motor function is impaired.

Pregnancy
Total parenteral nutrition
Gastric surgery ± truncal vagotomy
High spinal cord injury
somatostatinoma
Chronic octreotide treatment
Obesity

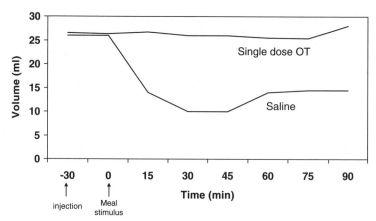

Graph 8.1 Effect of 50 μg Octreotide (OT) or saline injections, given 30 min before food, on meal stimulated gallbladder emptying, as judged by serial measurements of gallbladder volume in control subjects. Data taken from reference 15.

acromegalic patients will develop gallbladder stones.[14] Therefore, studies of acromegalic patients on long-term OT treatment provide a predictable and unique "model" of human gallstone formation.

It has been shown that OT virtually abolishes meal-stimulated gallbladder emptying in both control subjects and acromegalic patients.[15] Thus, when a lipid-rich liquid test meal is given some 30 min after an injection of either saline (placebo) or 50 μg OT, the normal emptying response of the gallbladder (studied by ultrasound) is abolished by the somatostatin analogue (Graph 8.1).

Initially it was assumed that the GB stasis seen in patients with somatostatinomas, or during OT treatment, was the sole mechanism for the development of OT-induced stones. However, it was also important to study changes in bile composition and physical chemistry. In order to study this, fresh gallbladder bile was obtained using the technique of ultrasound guided percutaneous, transhepatic fine needle puncture of the gallbladder[16] – a method pioneered in Germany by Swobodnik *et al.*[17]

Samples of gallbladder bile obtained in this way were analysed for bile acids, phospholipids and cholesterol (from which the cholesterol saturation indices were derived). Biliary bile-acid composition was then measured by HPLC. The vesicles were separated from micelles by sucrose density gradient ultra-centrifugation and the cholesterol microcrystal nucleation time measured as described above.

These analyses were carried out in three groups of patients:[18]

(i) a disease control group of nonacromegalic patients with conventional gallstone disease and two groups of acromegalic patients;
(ii) those with OT associated gallbladder stones and
(iii) those who had received no OT treatment and were gallstone-free by ultrasound

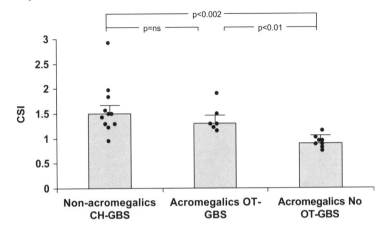

Graph 8.2 Biliary cholesterol saturation index (CSI) in nonacromegalic patients with cholesterol GBS (CH-GBS), acromegalic patients with Octreotide-associated GBS (OT-GBS), and stone-free patients before Octreotide treatment (No OT-GBS). The value of CSI = 1.0 indicates the limit of cholesterol solubility; points above the line are supersaturated, whereas those below the line are unsaturated with cholesterol. (All graphs show mean values ± SEMs.) Data taken from reference 18.

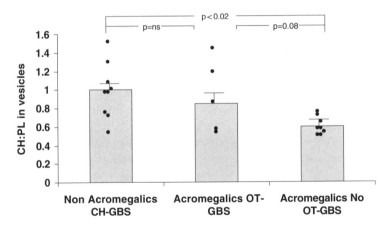

Graph 8.3 Group data for the cholesterol:phospholipid (CH/PL) molar ratio in the vesicles prepared from gallbladder bile. Data taken from reference 18.

The results of these studies,[18] showed that the acromegalic patients with iatrogenic gallstones all had supersaturated bile. This was apparently due to a relative excess of biliary cholesterol, represented by the cholesterol saturation index (CSI), and the excess cholesterol was found mainly in the vesicular fraction (Graph 8.2). The molar ratio of cholesterol:phospholipids in the vesicles was high, implying that they were unstable and vulnerable to cholesterol microcrystal precipitation (Graph 8.3). Indeed, as predicted, the cholesterol microcrystal nucleation time was abnormally rapid (Graph 8.4).

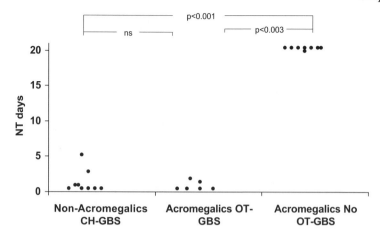

Graph 8.4 Group data for the cholesterol microcrystal nucleation time (NT) measures
in days. The value at 10 days represents the limit above which nucleation
time is normal. Data taken from reference 18.

This profile of changes was comparable to that seen in patients with "conventional" cholesterol gallstone disease, the only difference being that, as a result of OT, the chemistry probably developed in days or weeks, as opposed to months or years in the case of spontaneous gallstone formation.[19]

8.9   Biliary Phospholipid Fatty-Acid Composition

Arachadonic acid (AA) -rich phospholipids (AAPL) have also been incriminated in the pathogenesis of cholesterol gallbladder stones.[20] They may induce cholesterol supersaturation and nucleation of microcrystals in gallbladder bile in two ways: (i) by favouring the preferential transfer of AAPL from vesicles to micelles, as a result of which the vesicles become enriched with cholesterol and vulnerable to microcrystal precipitation, and (ii) by stimulating mucous glycoprotein (MGP) synthesis and secretion by the gallbladder mucosa.[21]

To study this, Pereira *et al.*,[22] measured the molecular species of phosphatidylcholine (PC – the principal phospholipid in bile) in gallstone patients. They found that the most abundant PC in bile had 16 carbons and no double bonds in the first (sn1) position, and 18 carbons with two unsaturated double bonds in the second (sn2) position. The second most common PC was (16:0–18:2), while the third was arachadonic acid-rich (16:0–20:4), which accounted for approximately 10–15% of the total biliary PC. The other AAPC (16:0–20:4) accounted for only 1% of the total. Nonetheless, these two arachadonic-acid-rich phospholipids were the only two molecular species of PC to change significantly during OT treatment.[22] Thus, the mean PC 16:0–20:4 increased from approximately 8% of the total before, to around 12% during OT therapy ($p < 0.01$) and, once again, this was associated with a significant increase in biliary cholesterol saturation. Moreover, there was a weak, but significant, linear relationship between the percentage AAPL and the proportion of DCA (% of total biliary bile acids) in bile.

8.10 Biliary Bile-Acid Composition

As indicated previously, deoxycholic acid (DCA) has long been implicated in the pathogenesis of cholesterol gallstones. The evidence for this was summarised in a major review article by Marcus and Heaton.[23] In brief, enrichment of bile with the hydrophobic bile acid, DCA, favours high biliary cholesterol secretion rates and supersaturated bile.[24] It has been shown that there are significant linear correlations between the %DCA in bile and (i) the mole % cholesterol and (ii) the cholesterol saturation indices.[25,26] Furthermore, in patients with abnormally rapid cholesterol crystal nucleation times (<5 days), the mean %DCA in bile is significantly higher than that in individuals who had abnormally long nucleation times (>10 days).[27]

When the proportion of biliary DCA, expressed as a percentage of total bile acids, was measured in the same three groups of individuals in the Octreotide studies, a similar pattern of results was found to that described above for bile lipids – namely, low values (approximately 12%) in the so-called controls and significantly higher values (approximately 24%) in the two groups of stone carriers.[18]

To see whether these changes in bile-acid composition were due to the OT treatment or to the resultant stones, the authors examined a small number of acromegalic patients whose bile composition was studied on two separate occasions, before and during OT treatment. Once again, the mean proportion of DCA in bile (percentage of total bile acids) doubled from approximately 12% before treatment to around 24% during OT therapy, independent of iatrogenic stone formation (Graph 8.5). At the same time, the cholesterol saturation index increased significantly and the bile changed from being unsaturated in cholesterol before treatment, to supersaturated as a result of the OT therapy.

8.11 The Adverse Effects of Increased Proportions of DCA in Bile on Cholesterol Gallstone Formation

There are multiple ways by which an increase in the percentage DCA in bile may pre-dispose to cholesterol gallstone formation. Carulli *et al.*,[24] showed that DCA-rich bile induces biliary cholesterol hyper-secretion when compared with other bile acids (Graph 8.6). This is likely to be due to the greater hydrophobicity and detergent effect of DCA, which would be able to solubilise the lipids in the canalicular cell wall more readily. It may well explain why there have been linear relationships demonstrated between the percentage DCA in bile and (i) the mole percentage cholesterol and (ii) the cholesterol saturation index in bile.[27,28] This may also explain why there is a link between the

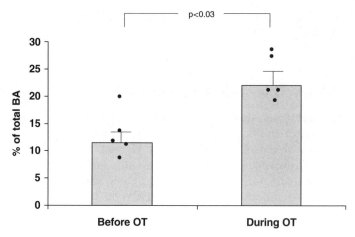

Graph 8.5   Paired data for the proportion of DCA (percentage of total bile acids) in bile before and during Octreotide treatment (300 μg/day for 8 months). Data taken from reference 18.

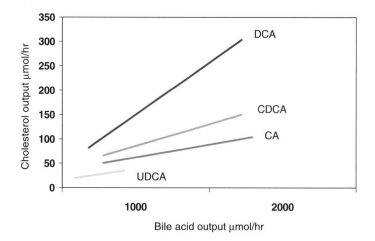

Graph 8.6   DCA, deoxycholic acid; CDCA, chenodeoxycholic acid; CA, cholic acid; UDCA, ursodeoxycholic acid. Data taken from reference 24.

percentage DCA in bile and the cholesterol microcrystal nucleation time.[27] As noted above, the percentage of DCA also correlates with the percentage arachadonic acid-rich phospholipids in bile.[22] In turn, this is linked both to mucous glycoprotein synthesis and secretion by the gallbladder, and to impaired gallbladder emptying.[21]

8.12 Mechanism for the Increase in the % Biliary DCA Conjugates in Bile during Octreotide Treatment

The results of several studies showed that, in nonacromegalic subjects, the mouth-to-caecum transit time (MCTT) is prolonged by a single 50-µg injection of OT.[29–31] However, in these studies there was no information about the effects of OT on MCTT in acromegalic patients, particularly those receiving long-term treatment as opposed to a single injection, and none about the effects of OT on large-bowel transit time (LBTT), which is of importance because the colon is the principal site for DCA formation.[32,33]

Hussaini *et al.*[34] confirmed that OT markedly prolongs MCTT, not only in control subjects but also in acromegalic patients. However, it took the power of paired studies,[35] to demonstrate that OT also significantly prolongs LBTT. In acromegalic patients studied before and during treatment, the mean colonic transit time increased by approximately 15 h ($p < 0.001$). At the same time, the percentage DCA in fasting serum (a surrogate marker for the percentage DCA in bile) also increased significantly. These two observations immediately provoked the question, what is the relationship between LBTT and the various markers of DCA metabolism?

8.13 The Relationship between Colonic Transit and DCA Metabolism

In 65 individuals (control subjects and acromegalic patients treated and untreated with OT) it has been shown that there was a significant linear relationship between total serum DCA (expressed as a percentage of serum total bile acids) and LBTT, in hours ($r = 0.70$; $p < 0.001$).[35] More importantly, these authors went on to show, in a smaller number of patients ($n = 32$), that there was an even stronger relationship ($r = 0.82$; $p < 0.001$) between LBTT and the DCA pool size (measured using serum sampling, gas chromatography-mass spectrometry GC-MS and stable isotope dilution).[36] (See Graph 8.7.)

Moreover, the conversion rate of ^{13}C-labelled cholic acid to ^{13}C-DCA was significantly greater during OT than before the somatostatin analogue treatment.

8.14 The Role of Prolonged Intestinal Transit in the Pathogenesis of Conventional Cholesterol Gallstones

Although the results of the studies in acromegalic patients with OT-induced gallstones are interesting, their importance would be limited if they were not also relevant to conventional gallstone disease (unrelated to acromegaly or OT).

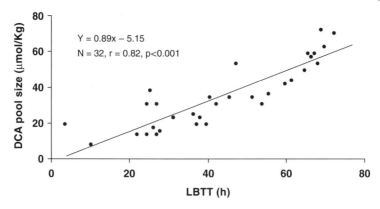

Graph 8.7   Relationship between large-bowel transit time (LBTT) and the DCA pool
size. Data taken from reference 36.

To date, the evidence that intestinal transit is also important in sporadic cho-
lelithiasis is limited, but that which is available all points in the same direction.

The first observation in this field was made by Heaton *et al.*[37] In normal-
weight women with gallstones but no other obvious risk factors, the Bristol
investigators found that, compared with age- and sex-matched controls, the
gallstone carriers had almost a 20-h longer whole gut transit time. Despite this,
their mean faecal wet weight was only half that of the controls. Put another
way, the women with gallstones had slow transit constipation.

Then in 1995, Shoda and colleagues from Japan and Sweden found that
gallstone carriers had significantly prolonged small-bowel transit times compared with controls.[38] They too had a two-fold increase in the percentage DCA
in bile that was associated with supersaturated bile, in contrast to the unsaturated bile found in the gallstone-free controls.

Similar findings have been reported by Azzaroli *et al.* from Bologna in
Italy.[39] In fact, these investigators measured oro-caecal transit time by three
different methods and again showed that the MCTT was significantly longer in
gallstone patients than in matched controls.

8.15   The Inter-Relationship between LBTT, Colonic
Anaerobic Bacteriology, Bile-Acid-Metabolising
Enzymes, Intraluminal pH and DCA Metabolism

In London, the Guy's Hospital group[40] wished to see if they could extrapolate
validly from OT-treated acromegalic patients to those with sporadic gallstone
disease. They also wished to study further the mechanism whereby prolongation of colonic transit might influence DCA metabolism, biliary cholesterol secretion and saturation, and therefore the risk of cholesterol gallstone
formation. To study this, Thomas *et al.*[40,41] developed a working hypothesis

based on three separate, but related, possible mechanisms. First, they argued that the prolonged colonic transit might allow more time for anaerobic bacterial proliferation in the colon and/or an increase in the bacterial enzymatic conversion of conjugated cholic acid into unconjugated DCA. Second, the results of previous studies had suggested that prolongation of colonic transit leads to higher than normal colonic luminal pH.[42,43] This, in turn, should favour increased solubilisation (and therefore increased bioavailability) of the newly formed DCA. Third, they reasoned that the increased LBTT might simply allow more time for DCA absorption from the colon (presumably by passive nonionic diffusion).

To test these ideas, the Guys' Hospital group[40] studied 20 "controls" who were stone-free by ultrasound, and 20 patients with presumed cholesterol-rich gallstones (as judged by maximum gallstone attenuation scores of <100 Hounsfield units on CT scanning, *in vivo*). All 40 individuals were undergoing clinically indicated left-sided colonoscopy, which had shown that they all had normal colons, both by endoscopy and at biopsy. In these 40 patients, LBTT was measured using radio-opaque marker shapes,[44] colonic luminal pH was measured by radiotelemetry,[45,46] while serum DCA was again measured by GC-MS.[47] Furthermore, in preparation for colonoscopy, the contents of the left colon were "washed out" with a phosphate enema, with the aim of leaving the right colon undisturbed. In this way, it was possible to examine the left colon endoscopically, after which the colonoscope was advanced, proximal to the hepatic flexure, so that aspirates of luminal contents from the caecum and ascending colon might be obtained. Then, in the colonic aspirates, the investigators studied quantitative bacteriology (using standard methods) and the activity of the bacterial enzymes responsible for bile-acid deconjugation and 7α-dehydroxylation.

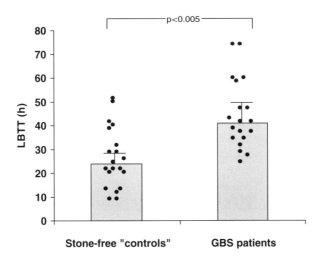

Graph 8.8   LBTT in stone-free control subjects and patients with cholesterol GBS. Data taken from references 40 and 41.

In brief, the results of these studies[40,41] showed that, compared with stone-free controls, the gallstone patients had significant increases in: (i) LBTT (Graph 8.8); (ii) total and Gram-positive anaerobes (Graph 8.9); (iii) cholic acid 7α-dehydroxylation (Graph 8.10); (iv) the luminal pH in both proximal and distal colon (Graph 8.11); and (v) the percentage DCA in fasting serum (and, by implication, the percentage DCA in bile) (Graph 8.12).

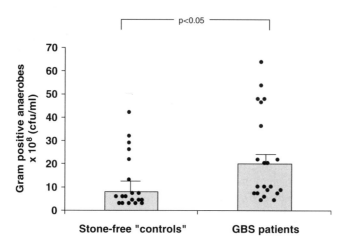

Graph 8.9   Gram-positive anaerobic bacterial counts measured in colony-forming units per millilitre of caecal aspirate from stone-free control subjects and patients with cholesterol GBS. Data taken from references 40 and 41.

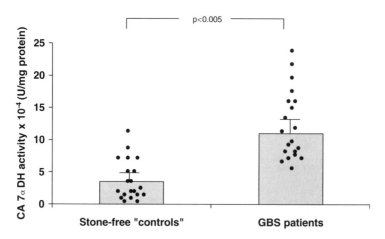

Graph 8.10 7α-dehydroxylase activity in stone-free controls and patients with cholesterol GBS. Data taken from references 40 and 41.

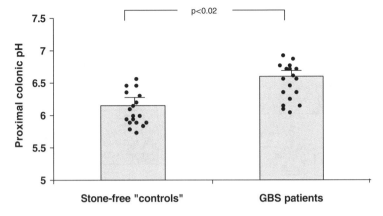

Graph 8.11 Proximal colonic pH using an expanded scale, from stone-free control subjects and patients with cholesterol GBS. Data taken from references 40 and 41.

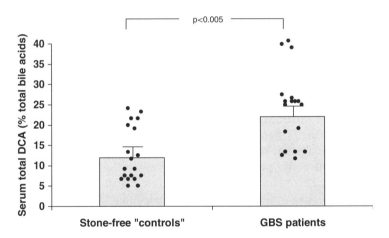

Graph 8.12 Proportion of DCA, expressed as a percentage of total serum bile acids, in stone-free control subjects and patients with cholesterol GBS. Data taken from references 40 and 41.

Given the large number of variables, Thomas *et al.*[41] argued that it was important to carry out both univariate and multi-variate analyses of the data. Both types of analysis showed that LBTT, the activity of the 7α-dehydroxylation, and the distal colonic luminal pH were all significant risk factors for high proportions of DCA in serum (and by implication, in bile). Based on these results, the authors concluded that there was indeed prolonged colonic transit in conventional gallstone patients and that this, in turn, was likely to result in increased DCA formation, solubilisation (and therefore bioavailability) and absorption.

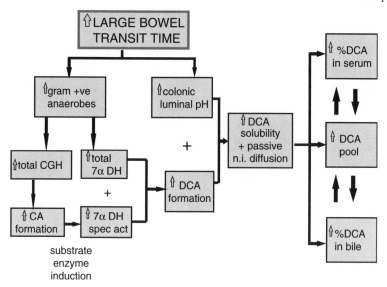

Figure 8.1   Flow diagram to represent proposed pathway by which prolonged large
bowel transit results in higher biliary and serum DCA levels.

8.16   Summary

The pathogenesis of cholesterol-rich gallstones is clearly multi-factorial. The
precipitation and nucleation of cholesterol microcrystals from supersaturated
bile remains a critical step in gallstone formation. The contribution of gallbladder stasis, including that associated with octreotide treatment has been
discussed, but also the physico-chemical changes that occur in octreotide-
treated patients. This "bad bile" is supersaturated with cholesterol, has excess
cholesterol in vesicles, rapid microcrystal nucleation times and an increased
percentage of DCA in bile. This seems to be associated with a prolongation of
large bowel transit time, and it has also been shown that patients with spontaneous gallstone disease also have prolonged large-bowel transit times. The
associated changes in colonic microbiology and bile-acid metabolising enzymes, along with higher intra-colonic pH values leads to increased DCA
formation, solubilisation and absorption. The resulting increased biliary DCA
levels leading to cholesterol gallstone formation.

This sequence of events is summarised in Figure 8.1.

References

1. W. Kratzer, R. A. Mason and V. Kachele, *J. Clin. Ultrasound*, 1999, **27**, 1–7.
2. G. D. Friedman, C. A. Raviola and B. Fireman, *J. Clin. Epidemiol.*, 1989,
   42, 127–136.
3. A. Caroli, G. Del Favero and F. Di Mario, *Gut*, 1992, **33**, 698–700.

4. W. H. Admirand and D. M. Small, *J. Clin. Invest.*, 1968, **47**, 1043–1052.
5. P. J. Thomas and A. F. Hofmann, *Gastroenterology*, 1973, **65**, 698–700.
6. A. Reuben, P. N. Maton, G. M. Murphy and R. H. Dowling, *Clin. Sci.*, 1985, **69**, 71–79.
7. L. J. Bennion and S. M. Grundy, *J. Clin. Invest.*, 1975, **56**, 996–1011.
8. Z. R. Vlacevic, C. C. Bell, I. Buhac, J. T. Farrar and L. Swell, *Gastroenterology*, 1979, **59**, 165–173.
9. E. A. Shaffer and D. M. Small, *J. Clin. Invest.*, 1977, **59**, 828–840.
10. G. J. Somjen and T. Gilat, *FEBS Lett.*, 1983, **156**, 265.
11. A. Kibe, A. C. Breuer, R. T. Holzbach, *Gastroenterology*, 1984, **86**, 1326 (Abstr).
12. K. R. Holan, R. T. Holzbach and R. E. Hermann, *Gastroenterology*, 1979, **77**, 611–617.
13. M. F. J. Stolk, K. J. van Erpecum and H. P. F. Koppeschaar, *Gut*, 1993, **34**, 808–813.
14. R. H. Dowling, S. H. Hussaini, G. M. Murphy, G. M. Besser and J. A. H. Wass, *Metabolism:Clinical and Experimental*, 1992, **41**(suppl. 2), 22–33.
15. S. H. Hussaini, S. P. Pereira, C. Kennedy, P. Jenkins, G. M. Murphy, J. A. H. Wass and R. H. Dowling, *Bile acids in Gastroenterology:Basic and Clinical Advances.*, Kluwer Academic Publishers, 1995, 319–330.
16. S. H. Hussaini, C. Kennedy, S. P. Pereira, J. A. H. Wass and R. H. Dowling, *Brit. J. Radiol.*, 1995, **68**, 271–276.
17. W. Swobodnik, N. Hagert, P. Janowitz and H. Wenk, *Radiology*, 1991, **178**, 755–758.
18. S. H. Hussaini, G. M. Murphy, C. Kennedy, G. M. Besser, J. A. H. Wass and R. H. Dowling, *Gastroenterology*, 1994, **107**, 1503–1513.
19. D. M. Small, *New Eng. J. Med.*, 1980, **302**, 1305–1307.
20. A. Cantafora, M. Angelico and A. DiBase, *Lipids*, 1981, **16**, 589–592.
21. M. C. Carey and M. J. Cahalane, *Gastroenterology*, 1993, **95**, 508–523.
22. R. H. Dowling, *Gut*, 1995, **36**(S1), A47(Abs).
23. S. N. Marcus and K. W. Heaton, *Gut*, 1988, **29**, 522–533.
24. N. Carulli, P. Loria, C. Bertolotti, M. Ponz de Leon, D. Menozzi, G. Medici and I. Piccagli, *J. Clin. Invest.*, 1985, **74**, 616–624.
25. S. H. Hussaini, S. P. Pereira, G. M. Murphy and R. H. Dowling, *Gut*, 1995, **35**(Suppl 3), A39.
26. A. F. Hofmann, S. M. Grundy and J. M. Lachin, *Gastroenterology*, 1982, **83**, 738–752.
27. S. H. Hussaini, S. P. Pereira, G. M. Murphy and R. H. Dowling, *Hepatology*, 1995, **22**, 1735–1742.
28. D. Alvaro, F. Angelico and A. F. Atilli, *Biochim. Biophys. Acta.*, 1986, **45**, 761–768.
29. H. S. Fuessl, G. Carolan, G. Williams and S. R. Bloom, *Digestion*, 1987, **36**, 101–107.
30. N. Moller, G. Petrany and D. Cassidy, *Clin. Sci.*, 1988, **75**, 345–350.
31. L. J. D. O'Donnell, A. J. M. Watson, D. Cameron and M. J. G. Farthing, *Aliment Pharmacol. Ther.*, 1990, **4**, 177–182.

32. M. J. Hill and B. S. Drasar, *Gut*, 1974, **6**, 318–323.
33. T. Midtvedt, *Am. J. Clin. Nutr.*, 1974, **27**, 1341–1347.
34. S. H. Hussaini, S. P. Pereira and M. J. Veysey, *Gut*, 1996, **38**, 775–783.
35. M. J. Veysey, L. A. Thomas and A. I. Mallet, *Gut*, 1999, **44**, 675–681.
36. M. J. Veysey, L. A. Thomas, A. Mallet, P. J. Jenkins, G. M. Besser, G. M. Murphy and R. H. Dowling, *Gastroenterology*, 2001, **121**, 812–822.
37. K. W. Heaton, P. M. Emmett, C. L. Symes and F. E. M. Braddon, *Lancet*, 1993, **341**, 8–10.
38. J. Shoda, B.-F. He and N. Tanaka, *Hepatology*, 1995, **21**, 1291–1302.
39. F. Azzaroli, G. Mazzella and P. Mazleo, *Am. J. Gastro.*, 1999, **94**, 2453–2459.
40. L. A. Thomas, M. J. Veysey, G. M. Murphy, R. H. Dowling, A. King and G. R. French, *Gastroenterology*, 2000, **119**, 806–815.
41. L. A. Thomas, T. Bathgate and M. J. Veysey, In: *Bile Acids and Cholestasis.*, eds. G. Paumgartner, A. Stiel, W. Gerok, D. Kepper, U. Leuschner, Kluwer Academic Publishers, Dordecht, Boston, London, 1999, 284.
42. L. El Oufir, B. Flourie and S. Bruley des Varannes, *Gut*, 1996, **38**, 870–877.
43. S. J. Lewis and K. W. Heaton, *Gut*, 1997, **41**, 245–251.
44. A. M. Metcalf, S. F. Phillips, A. R. Zinsmeister, R. L. MacCarty, R. W. Beart and B. G. Wolff, *Gastroenterology*, 1987, **92**, 40–47.
45. D. F. Evans, G. Pye and R. Bramley, *Gut*, 1988, **29**, 1035–1041.
46. J. Fallingborg, L. A. Christensen and M. Ingeman-Nielson, *Aliment Pharmacol. Ther.*, 1989, **3**, 605–613.
47. J. Goto, K. Watanabe and H. Miura, *J. Chromatogr.*, 1987, **388**, 379–387.

Subject Index

Note: Page numbers in *italic* refer to diagrams and tables. Page numbers in **bold** indicate more comprehensive treatment.